I0038116

CHANGING CHANNELS

Regional Information
for Developing Multi-benefit
Flood Control Channels
at the Bay Interface

PRIMARY AUTHORS
Scott Dusterhoff
Sarah Pearce
Lester McKee
Julie Beagle
Carolyn Doehring
Katie McKnight
Robin Grossinger

DESIGN AND LAYOUT
Ruth Askevold

APRIL 2017

PREPARED BY
San Francisco Estuary Institute-Aquatic Science Center

IN COOPERATION WITH
San Francisco Estuary Partnership
San Francisco Bay Conservation and Development Commission
San Francisco Bay Joint Venture

FUNDED BY
San Francisco Bay Water Quality Improvement Fund, EPA Region IX

SFEI PUBLICATION #801

SFEI | AQUATIC SCIENCE CENTER
SAN FRANCISCO ESTUARY INSTITUTE & THE AQUATIC SCIENCE CENTER

A PRODUCT OF FLOOD CONTROL 2.0

FLOOD 2.0
CONTROL

SUGGESTED CITATION

San Francisco Estuary Institute-Aquatic Science Center. 2017. *Changing Channels: Regional Information for Developing Multi-benefit Flood Control Channels at the Bay Interface.* A SFEI-ASC Resilient Landscape Program report developed in cooperation with the Flood Control 2.0 Regional Science Advisors, Publication #801, San Francisco Estuary Institute-Aquatic Science Center, Richmond, CA.

Version 1.1, May 2017 (reflects minor revisions to v1.0)

REPORT AVAILABILITY

Report is available at floodcontrol.sfei.org

COVER

Landsat 8 satellite imagery of the San Francisco Bay taken on April 16, 2013. This image shows a pulse of sediment entering the Bay from the surrounding lands, transported by creeks after a series of spring rain storms.

Imagery courtesy of U.S. Geological Survey and NASA.

CONTENTS

ACKNOWLEDGEMENTS

The Flood Control 2.0 project was funded by the EPA Region IX's San Francisco Bay Water Quality Improvement Fund. We give special thanks to Luisa Valiela, the EPA project manager, for all of her support and enthusiasm throughout the project. The work presented here benefited greatly from the insight and guidance provided by our science advisors. We thank the Regional Science Advisors for their participation in several workshops and their review of our technical products. The Regional Science Advisors were Peter Baye (Ecological Consultant), Letitia Grenier (SFEI, formerly Bayland Goals Project), Jeff Haltiner (Hydrology Consultant), Barry Hecht (Balance Hydrologics), Rob Leidy (EPA), Jeremy Lowe (SFEI, formerly ESA), Leonard Sklar (San Francisco State University), and Andy Gunther (Bay Area Ecosystems Climate Change Consortium). We also thank the National Science Advisors for providing external perspective and helping hone the technical and management aspects of this work. The National Science Advisors were Pinar Balci (NYC Department of Environmental Protection), Derek Booth (University of California Santa Barbara), Doug Shields (cbec, inc.), and Si Simenstad (University of Washington).

This work would not have been possible without all of the information we received from Bay Area flood control agencies, counties, municipalities, agencies, and consultants. We thank the following people for giving their time and helping with data transfer: Paul Detjens (Contra Costa County Flood Control and Water Conservation District); Sara Duckler, Ray Fields, Shree Dharasker, and Scott Katric (Santa Clara Valley Water District); Roger Leventhal, Hannah Lee, Neal Conaster, Hugh Davis, Pat Valderama, and Joanna Dixon (Marin County Department of Public Works); Rohin Saleh, Moses Tsang, and Arthur Valderrama (Alameda County Flood Control District); Jon Niehaus (Sonoma County Water Agency); Jeremy Sarrow (Napa County Flood Control and Water Conservation District); Tim Tucker and Joe Enke (City of Martinez); Rich Walkling (Restoration Design Group); Jeff Brown (City of Hercules); Leticia Alvarez (City of Belmont); Julie Casagrande, Mark Chow, and Carole Foster (San Mateo County Department of Public Works); Nixon Lam (San Francisco International Airport); Kevin Murray (San Francisquito Creek Joint Powers Authority); Pam Tuft (City of Petaluma); and Jessica Burton Evans and Shelah Swett (US Army Corps of Engineers).

We are grateful to the Flood Control 2.0 project core team and the project partners for all of their invaluable contributions to the work presented in this report. The core team was Caitlin Sweeney and Adrien Baudrimont (San Francisco Estuary Partnership); Brenda Goeden, Pascale Soumoy, and Anniken Lydon (San Francisco Bay Conservation and Development Commission); and Beth Huning and Sandra Scoggin (San Francisco Bay Joint Venture). The project partners included Len Materman and Kevin Murray (San Francisquito Creek Joint Powers Authority); Liz Lewis, Laurie Williams, and Roger Leventhal (Marin County Department of Public Works); and Paul Detjens, Mike Carlson, and Mitch Avalon (Contra Costa County Flood Control and Water Conservation District). We are also indebted to Carl Morrison and the Bay Area Flood Protection Agencies Association (BAFPAA) for helping conceive this project and supporting it from start to finish.

We are also grateful to the SFEI staff members and partners who helped with this report and many other aspects of the Flood Control 2.0 project: Micha Salomon, Sean Baumgarten, Erin Beller, April Robinson, Lawrence Sim, Steve Hagerty, Jen Hunt, Josh Collins, Laurel Collins (Watershed Sciences), and Meredith Williams (California Department of Toxic Substance Control).

Finally, we thank Luisa Valiela, Jeff Haltiner, Jeremy Lowe, Si Simenstad, and Roger Leventhal for helpful feedback on an early draft of this report.

EXECUTIVE SUMMARY

Over the past 200 years, many of the channels that drain to San Francisco Bay have been modified for land reclamation and flood management. The local agencies that oversee these channels are seeking new management approaches that provide multiple benefits and promote landscape resilience. This includes channel redesign to improve natural sediment transport to downstream bayland habitats and beneficial re-use of dredged sediment for building and sustaining baylands as sea level continues to rise under a changing climate. Flood Control 2.0 is a regional project that was created to help develop innovative approaches for integrating habitat improvement and resilience into flood risk management at the Bay interface. Through a series of technical, economic, and regulatory analyses, the project addresses some of the major elements associated with multi-benefit channel design and management at the Bay interface and provides critical information that can be used by the management and restoration communities to develop long-term solutions that benefit people and wildlife.

This Flood Control 2.0 report provides a regional analysis of morphologic change and sediment dynamics in flood control channels at the Bay interface, and multi-benefit management concepts aimed at bringing habitat restoration into flood risk management. The findings presented here are built on a synthesis of historical and contemporary data that included input from Flood Control 2.0 project scientists, project partners, and science advisors. The results and recommendations, summarized below, will help operationalize many of the recommendations put forth in the Baylands Ecosystem Habitat Goals Science Update (Goals Project 2015) and support better alignment of management and restoration communities on multi-benefit bayland management approaches.

Channel morphologic change at the Bay interface

In the mid-19th century, the locations where creeks met the tidal environment (i.e., fluvial-tidal [F-T] interfaces) were of four main types: creek connection directly to the Bay, creek connection to a tidal channel network, creek drained onto a tidal marshland, and creek unconnected to the tidal environment (except during large floods). The major drivers controlling the historical F-T interface type include stream power during floods and watershed sediment supply. Over the past 200 years, most channels have been altered to have a permanent connection to a tidal channel that flows through diked baylands or bay fill, or have been routed underground or filled in completely. Many channels that once connected naturally to tidal marshlands or a tidal channel and are now constrained by levees still have the physical characteristics (e.g., geomorphic setting and sediment load) conducive for moving freshwater and sediment out to the Bay and helping build and maintain bayland habitats.

Watershed sediment yield and sediment removal in flood control channels at the Bay interface

Current estimated average annual watershed sediment yields to the 33 major flood control channels at the Bay interface range from <100 to >1,500 tons/mi^2/yr and reflect differences in watershed geology, climate and land management factors. Over the past four decades, approximately two-thirds of the 5.8 million cubic yards of sediment removed from the flood control channels came from tidal reaches downstream of head of tide. Most of the sediment came from channels that were dredged, on average, at least once every five years, and most was taken to landfills or disposed of as a waste product. Sediment removal from these channels has cost $115M, with individual channel costs ranging from <$1,500 to >$5,000,000/mi^2 of channel dredged/yr. The sediment that passes though and is removed from these channels is a valuable commodity for baylands restoration. Redesign projects at the Bay interface should therefore prioritize approaches for moving sediment to support natural accretion of marsh plains and for building or nourishing coarser grained tidal landscape features (e.g., depositional fans and beaches).

Multi-benefit management measures for flood channels at the Bay interface

The findings from the F-T interface assessment and sediment analysis were combined with additional land use information to highlight potential opportunities for improving sediment delivery to baylands through natural and mechanical means. Of the 33 major flood control channels considered, 25 have the potential for reconnecting the channel to bayland habitats due to the presence of undeveloped land that could be flooded by river and tidal waters. For the other eight channels, beneficial sediment re-use appears to be the most viable option for getting sediment to bayland habitats due the lack of opportunities for channel reconnection. The physical setting of five South Bay creeks considered suggests consideration of a hybrid approach that includes both creek reconnection and beneficial re-use of sediment removed from inland reaches near the channels' historical terminus. The feasibility of these management actions would need to be determined through detailed constraints assessments and technical analyses.

Key recommendations for future work

More information is needed on the quantity and quality of sediment that deposits in and travels through flood control channels at the Bay interface. This information would ultimately help us develop resilient tidal habitat restoration projects and improve multi-benefit management approaches. Based on the findings from this report and other related efforts, we developed the following recommendations for data collection efforts and quantitative analysis focused on sediment dynamics:

Watershed sediment supply Continuous suspended load and episodic bedload data should be collected for at least the 33 major flood control channels discussed in this report and made publicly available.

In-channel sediment storage Channel cross-section surveys, longitudinal profile surveys, and sediment grain size analyses should be conducted regularly for at least the 33 major flood control channels discussed in this report and the results should be made publicly available.

Sediment removal Information from all sediment removal events (e.g., removal location, sediment volume, sediment grain size, sediment fate, and costs) should be entered into a publicly available database.

Future conditions Detailed numerical modeling of the impacts of climate change on sediment transport and deposition should be conducted for at least the major 33 flood control channels discussed in this report.

Additional sediment sources Bay sediment source analyses should be extended to include the many hundreds of stormwater pipe outfalls at the Bay interface.

San Francisquito Creek channel, July, 2013. (SFEI)

INTRODUCTION

1

BACKGROUND

Over the past two centuries, many of the rivers and streams that drain to San Francisco Bay have been modified for water supply, land reclamation, and flood risk management. These modifications include the installation of dams, building of concrete trapezoidal channels, channel realignment, and leveeing of channels and reclaiming historical floodplains and baylands (i.e., mudflats, tidal marshes, tidal-terrestrial transition zones). In many instances, these actions have had considerable impacts on the way in which water and sediment move through these channels and out to the Bay. Historically, many creeks transported watershed-derived sediment to baylands and out the Bay during wintertime storms. Now, many of these creeks are confined by flood control levees in their tidal reaches, which has resulted in decreased sediment supply to baylands, excess in-channel sedimentation, channel dredging to maintain flow capacity, and subsequent adverse dredging impacts to resident plants and wildlife. These modifications have also impacted the ecological functions provided by local creeks, including fish rearing and riparian habitat functions.

The local agencies that operate and maintain the flood control channels that drain to the Bay are coming under increasing pressure from regulatory agencies to manage or redesign flood infrastructure to provide beneficial uses beyond flood conveyance, including supporting the natural processes that create and maintain tidal marshlands. As such, flood control engineers and managers are beginning to focus on ways of getting sediment from flood control channels to downstream bayland habitats through re-establishing natural transport processes, where possible, and mechanical placement of dredged sediment where dredging must continue to maintain the required level of flood protection. Therefore, sediment being delivered to (and trapped in) these flood channels is now being seen as a valuable commodity for bayland restoration in the near-term and bayland survival over the long-term under a rising sea-level and a projected decrease in Bay suspended sediment concentrations (Goals Project 2015).

To help improve the ecological health and resilience of San Francisco Bay habitats, the EPA and other entities provided funding for a regional project called Flood Control 2.0. The main goal of the project was to develop information that is useful for integrating habitat improvement and bayland resilience into flood risk management in intertidal flood control channels around San Francisco Bay. The project outputs are a series of management "tools" that include the following:

- a regional analysis of historical and contemporary channel morphology, contemporary watershed sediment delivery and in-channel sedimentation dynamics, and management concepts aimed at improving physical and ecological functioning at the Bay interface

- historical ecology analyses for the tidal portion of selected channels highlighting habitat changes since the beginning of intensive European-American settlement

- multi-benefit landscape "visions" at the Bay interface for selected channels that draw on historical ecology analyses and provide a suite of potential management actions that incorporate process-based habitat restoration into flood risk management and could ultimately lead to a more functional and resilient landscape

Pacheco Marsh (right). (SFEI)

- benefit-cost analyses that provide quantitative comparisons of the benefit-cost ratios associated with the landscape vision actions to those associated with maintaining the current management approach into the future (e.g., channel dredging and levee maintenance)

- regulatory guidance documents that provide examples of regulatory issues associated with previous flood control projects at the Bay interface and recommendations for regulatory improvements that could facilitate better project planning and implementation

- an online "marketplace" called SediMatch where sediment removed from flood control channels is advertised to the bayland management and restoration communities

- interviews with Flood Control 2.0 project partners detailing the issues associated with operating and maintaining flood controls, the channel management challenges caused by a changing climate, and the role that the Flood Control 2.0 project can play in highlighting opportunities for multi-benefit management approaches

These project outputs are intended to help flood control agencies and their partners design landscapes to promote improved sediment transport through flood control channels, improved flood conveyance, and the restoration of resilient bayland habitats. In combination with other regional plans, this project provides an overarching framework to flood control managers and the restoration community for planning sustainable, long-term, multi-benefit redesign projects in the context of infrastructure, regulatory, and economic challenges.

THE FLOOD CONTROL 2.0 PROJECT TOOLBOX CAN BE FOUND AT **floodcontrol.sfei.org.**

FLOOD CONTROL 2.0 PROJECT STRUCTURE

PROJECT LEADS

- SAN FRANCISCO ESTUARY PARTNERSHIP (SFEP)
- SAN FRANCISCO ESTUARY INSTITUTE (SFEI)
- SAN FRANCISCO BAY CONSERVATION AND DEVELOPMENT COMMISSION (BCDC)
- SAN FRANCISCO BAY JOINT VENTURE (SFBJV)

PROJECT PARTNERS

- CONTRA COSTA COUNTY FLOOD CONTROL AND WATER CONSERVATION DISTRICT (CCCFCWCD)
- MARIN COUNTY DEPARTMENT OF PUBLIC WORKS (MCDPW)
- SAN FRANCISQUITO CREEK JOINT POWERS AUTHORITY (SFCJPA)

REGIONAL SCIENCE ADVISORY TEAM

- PETER BAYE (ECOLOGICAL CONSULTANT)
- LETITIA GRENIER (SFEI, FORMERLY BAYLAND GOALS PROJECT)
- JEFF HALTINER (HYDROLOGY CONSULTANT)
- BARRY HECHT (BALANCE HYDROLOGICS, INC.)
- ROB LEIDY (US EPA)
- JEREMY LOWE (SFEI, FORMERLY ESA)
- LEONARD SKLAR (SAN FRANCISCO STATE UNIV.)

NATIONAL SCIENCE ADVISORY TEAM

- PINAR BALCI (NYC DEPT. OF ENVIRONMENTAL PROTECTION)
- DEREK BOOTH (UNIV. OF CALIFORNIA SANTA BARBARA)
- DOUG SHIELDS (cbec, INC.)
- SI SIMENSTAD (UNIV. OF WASHINGTON)

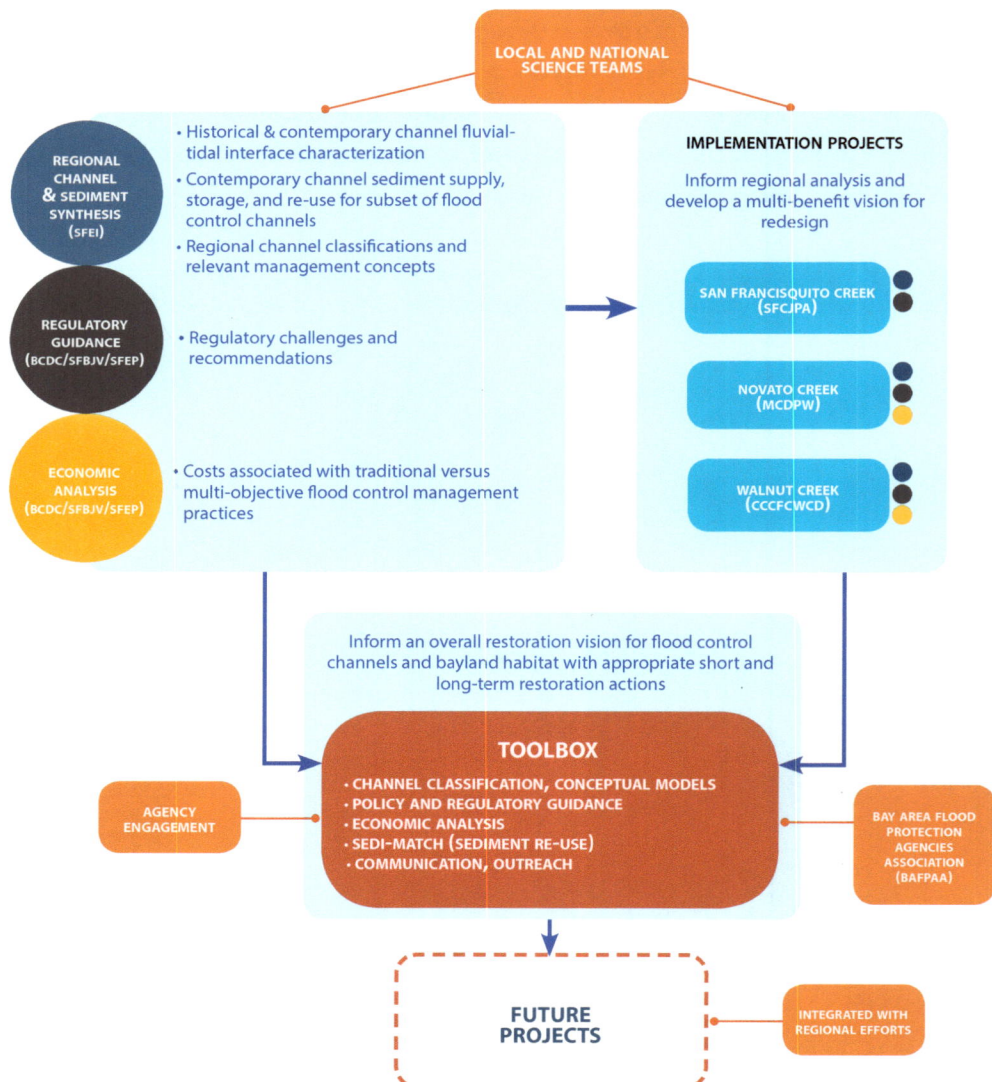

REGIONAL ANALYSIS APPROACH AND COMPONENTS

This report provides a regional analysis of morphologic change and sediment dynamics in flood control channels at the Bay interface, and multi-benefit management concepts aimed at bringing habitat restoration into flood risk management. The goal of the synthesis is to provide the local management and restoration communities with information that can be used to develop concepts early in the channel redesign process based on opportunities that could exist for bayland habitat creation and maintenance. The information in this report will help operationalize many of the management recommendations put forth in the recently released Baylands Ecosystem Habitat Goals Science Update (Goals Project 2015) and support better alignment of bayland management and restoration communities on approaches that help meet flood risk management needs while restoring resilient tidal landscapes.

This report is divided into chapters that provide the following information:

- *Chapter 2: Channel morphologic change at the Bay interface* – This chapter provides a high level overview of morphologic change where watershed channels drain to the Bay (i.e., at the Bay interface or fluvial-tidal interface) since the mid-19th century. The analysis presented provides an indication of the magnitude of channel change and associated habitat change around the Bay, as well as an indication of primary landscape drivers for current in-channel sediment deposition issues.

- *Chapter 3: Watershed sediment yield and sediment removal in flood control channels at the Bay interface* – This chapter provides a detailed assessment of contemporary watershed sediment delivery to and sediment removal from major flood control channels draining the largest watersheds in the region. The analysis quantifies the volume of sediment removed over the past several decades and the portion removed just upstream and within the tidal portion of flood control channels, which helps clarify the primary drivers for excess sediment accumulation, as well as the amount of sediment potentially available as a resource.

- *Chapter 4: Multi-benefit management measures for flood control channels at the Bay interface* – This chapter synthesizes the information presented in Chapters 2 and 3 and other related landscape information into high level management concepts (or measures) for the major flood control channels that focus on opportunities for improving sediment delivery to bayland habitats

Baylands Ecosystem Habitat Goals Science Update

RECOMMENDATIONS

1. *Restore estuary-watershed connections*

2. *Design complexity and connectivity into the Baylands*

3. *Restore and protect complete tidal wetlands systems*

4. *Restore the Baylands to full tidal action before 2030*

5. *Plan for the Baylands to migrate*

6. *Actively recover, protect and monitor wildlife*

7. *Develop and implement a comprehensive regional sediment-management plan*

8. *Invest in planning, policy, research, and monitoring*

9. *Develop a regional transition zone assessment program*

10. *Improve carbon management in the Baylands*

(both natural and mechanical) within the context of improving flood risk management. The process for assigning appropriate management concepts is described for selected example channels.

- **Chapter 5: Recommendations** – This chapter provides recommendations for future work to help implement management measures developed in Chapter 4, including additional data compilation and assessments in the contributing watershed and at the Bay interface.

The information and management recommendations presented in this report were used as the starting point for developing multi-benefit landscape "visions" for two flood control channels: lower Novato Creek in Marin County and lower Walnut Creek in Contra Costa County (see SFEI-ASC 2015 and SFEI-ASC 2016). Using an understanding of each creek's changes and contemporary sediment delivery and deposition dynamics, Flood Control 2.0 project scientists worked with flood control managers and a team of science advisors to develop a suite of management concepts aimed at bringing habitat restoration, maintenance, and resilience into flood risk management. These visions provide flood control agencies and their local partners with a roadmap for a more functional and resilient landscape that could be achieved through a series of coordinated projects over multiple years. This process would need to be supported by strong partnerships among local land owners and stakeholders, close coordination with the regulatory agencies overseeing permitting, and the necessary financial resources.

Pickleweed near the mouth of Walnut Creek, August, 2015. (SFEI)

CHANNEL MORPHOLOGIC CHANGE AT THE BAY INTERFACE

2

Detail, San Francisco Bay, USGS, 1915. (Courtesy of David Rumsey)

INTRODUCTION

The locations where fluvial (or riverine) environments meet tidal environments (i.e., fluvial-tidal [F-T] interfaces) are key delivery points of freshwater, sediment, and nutrients to the Bay. Historically, F-T interface settings around San Francisco Bay reflected a wide range of landscape conditions and characteristics including geology, topography, and hydrology. Over the past 200 years, water and sediment delivery to the F-T interface have been altered and the interface locations themselves have been extensively modified for the sake of land reclamation and flood control, resulting in channel sedimentation issues and impaired ecosystem functioning, and decreased resilience to climate change (Goals Project 2015). In this chapter, we describe the historical F-T interface types and the geomorphic drivers controlling their occurrence, and chronicle how F-T interfaces around the Bay have changed since the onset of intensive European-American settlement in the mid-19th century. Understanding the change from past to present conditions can help clarify the causes for current sediment deposition, as the underlying physical characteristics that help control sedimentation (e.g., landscape slope, bedrock geology, and watershed size) remain unchanged. This understanding can in turn help inform future management actions that improve sediment transport and work in concert with natural processes.

CONCEPTUAL UNDERSTANDING OF HISTORICAL AND CONTEMPORARY CHANNEL DYNAMICS

Historically, fluvial channels draining the watersheds surrounding San Francisco Bay would either terminate before reaching the Bay or discharge into the Bay, depending on the channel's ability to transport flow and sediment during flood events. Channels that ended inland would often terminate on alluvial fans at the mouths of steep canyons or on adjacent alluvial plains, although many had a connection to the Bay during extreme flood events. Channels that regularly made it to the Bay would deliver freshwater and sediment to the F-T interface and often downstream to tidal channel networks that extended through tidal marsh plains. Tidal channel networks were maintained over time by tidal scour, which is directly related to the size of the marsh and associated tidal prism (or the volume of water that enters and exits the marsh over a tidal cycle). Larger marsh plains had larger tidal prisms and scouring tidal flows that maintained a relatively deep and wide mainstem channel and a strong salinity gradient between the channel's tidal reach and upstream fluvial reach. The channels draining watersheds with very high sediment yields had natural levees at the F-T interface that extended into the marsh plain and supported a mosaic of riparian vegetation.

Today, most of the channels around the Bay are much different than they were in the mid-19th century. To allow for development and agriculture, channels were modified to move floodwaters and sediment quickly out to the Bay. This included enlarging and rerouting channels, extending disconnected channels to have a permanent connection to the Bay, and building networks of flood control levees along channels. These modifications have resulted in in-channel sedimentation and flood conveyance issues in many channels that often require regular maintenance. For example, some rerouted channels have excess watershed sediment deposition at unnatural channel bend locations where flood waters lose velocity. In addition, levee construction along tidal channels has resulted in the cessation of regular inundation of adjacent tidal marsh plains and reduced tidal prism, thereby causing excess sediment deposition and overall channel in-filling. As sea-level continues to rise, the location of the F-T interface will migrate inland, possibly causing new flooding risks associated with backwater effects and excess sediment deposition as fluvial flood waters meet higher tidal waters.

HISTORICAL LANDSCAPE

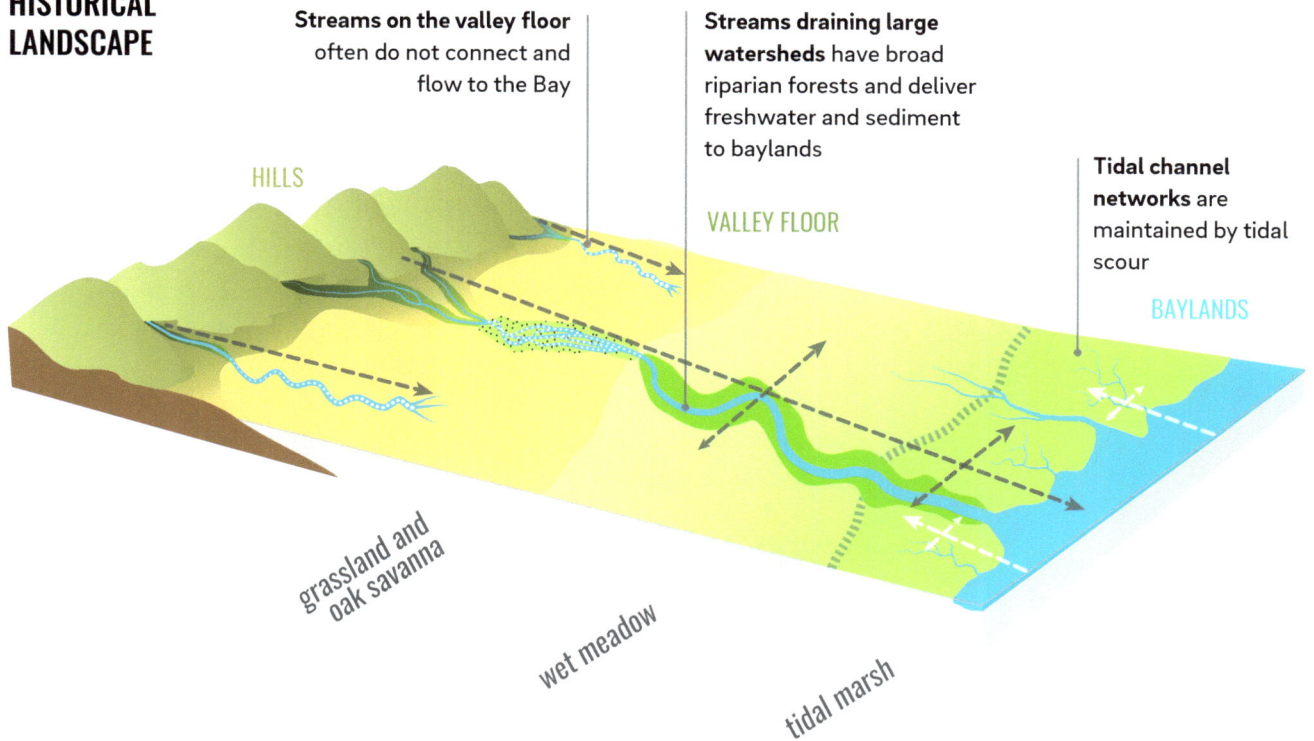

Streams on the valley floor often do not connect and flow to the Bay

Streams draining large watersheds have broad riparian forests and deliver freshwater and sediment to baylands

Tidal channel networks are maintained by tidal scour

HILLS

VALLEY FLOOR

BAYLANDS

grassland and oak savanna

wet meadow

tidal marsh

MODERN LANDSCAPE

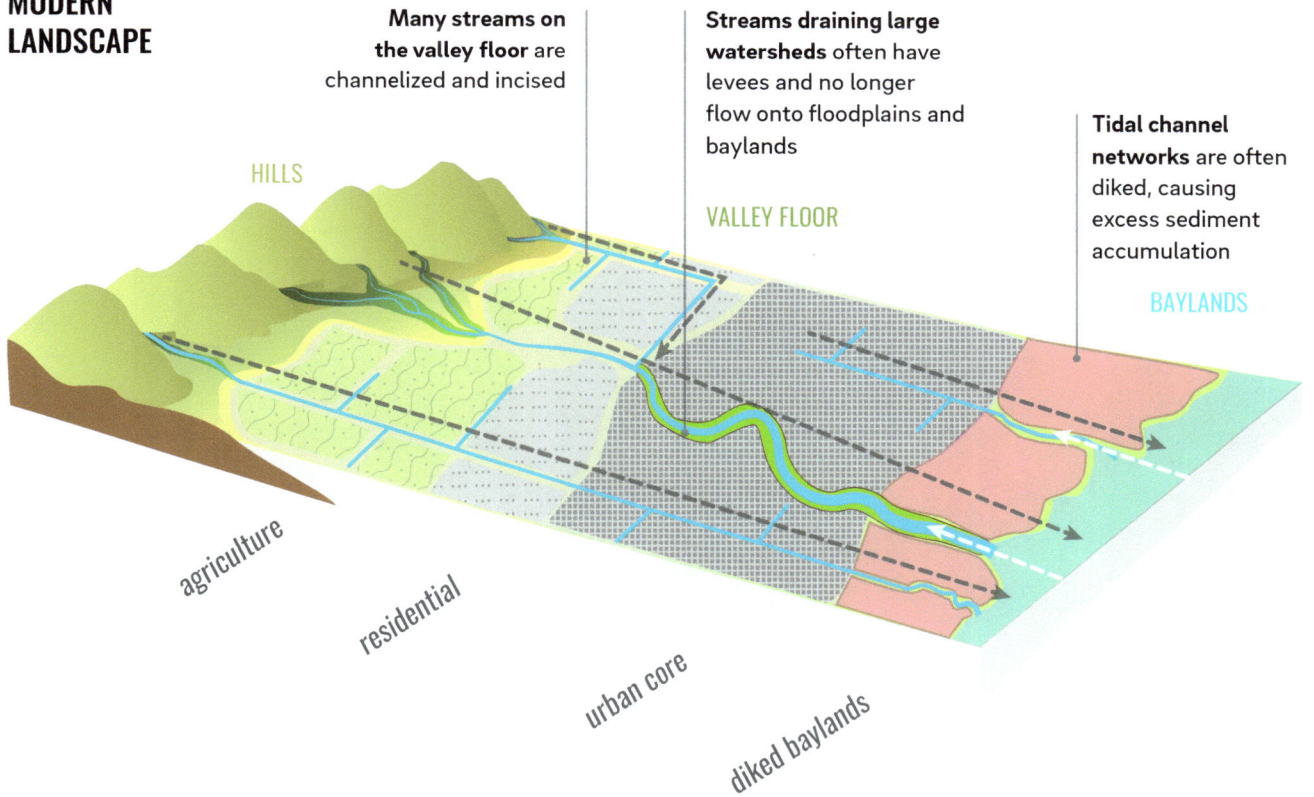

Many streams on the valley floor are channelized and incised

Streams draining large watersheds often have levees and no longer flow onto floodplains and baylands

Tidal channel networks are often diked, causing excess sediment accumulation

HILLS

VALLEY FLOOR

BAYLANDS

agriculture

residential

urban core

diked baylands

HYDROLOGY

Intermittent

Perennial

PHYSICAL PROCESS

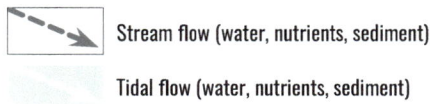

Stream flow (water, nutrients, sediment)

Tidal flow (water, nutrients, sediment)

Conceptual model of historical and contemporary dynamics of many channels draining to San Francisco Bay. Graphic developed in coordination with the Santa Clara Valley Water District.

METHODS

A range of sources were used to determine historical (mid-19th century) and contemporary channel F-T interface types around San Francisco Bay. For classification of the historical F-T interface types, we used U.S. Coastal and Geodetic Survey topographic sheets (or t-sheets) and existing historical ecology studies that used t-sheets as well as U.S. Geological Survey maps, U.S. Department of Agriculture soil surveys, Mexican land grant maps, General Land Office public land surveys, and a variety of historical documents. For contemporary F-T interface classification, we used the Modern Baylands GIS layer (SFEI 1998) and the Bay Area Aquatic Resources Inventory (BAARI) (SFEI 2014).

Channels were classified into three types, based on the nature of their pre-modification character and subsequent alteration:

- Channels that were historically connected to tidal marshes (or marshlands), a tidal marsh channel, or directly to the Bay through a discrete channel and currently have a connection to the tidal environment.

- Channels that were historically connected to tidal marshlands, a tidal marsh channel, or directly to the Bay through a discrete channel and have been routed underground or filled in.

- Channels that were historically disconnected from the tidal environment (except during extreme flows) but currently have a permanent connection through a discrete channel.

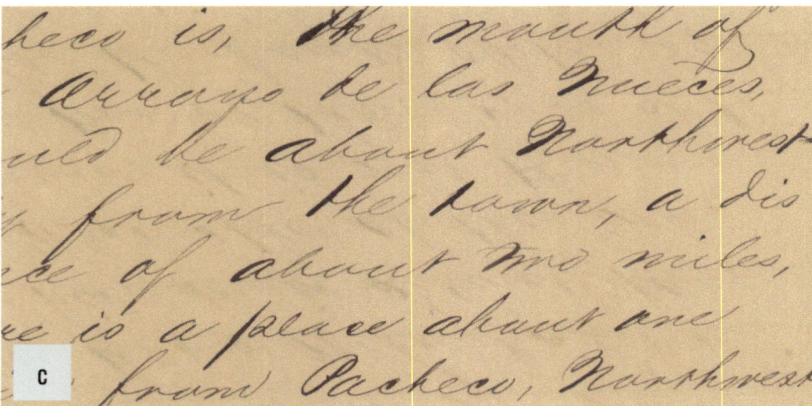

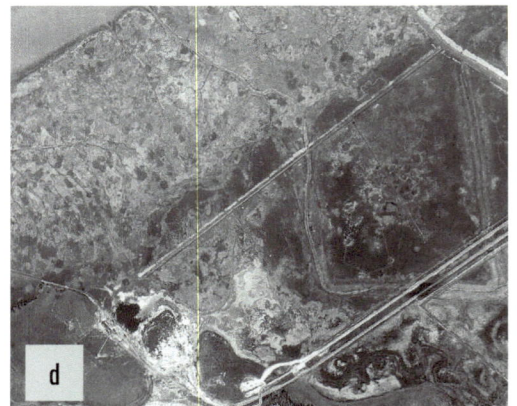

The historical F-T interface was defined as the mapped location at which a riverine channel first intersects with either the tidal marshlands, tidal marsh channel, the Bay, or at a channel's inland terminus for unconnected or disconnected channels. Because the level of detail for the historical marsh plain mapping is not the same for all channels (i.e., some channels have detailed historical ecology studies), the historical F-T interface designations are considered a first approximation. Controls on F-T interface type were examined through an assessment of local geology (Jennings et al. 1977 and Witter et al. 2006) and physical landscape characteristics such as watershed area, watershed topography and coastline morphology (derived from the USGS 10-m DEM).

The contemporary F-T interface was defined by the mapped location where present-day channels intersect with the historical tidal marshlands extent or merge with another historical channel. Contemporary channels that intersect with the historical marsh plain extent were further classified by determining the dominant land use adjacent to the channel before it reaches the Bay. The land type was determined using the Modern Baylands GIS layer (SFEI 1998) and BAARI (SFEI 2014), which designate land in the historical marshlands extent as diked baylands (i.e., managed areas isolated from the tides by dikes or levees), bay fill (i.e., areas where fine sediment has been placed to increase elevation and allow for development), and tidal marsh. Although the dominant land type was chosen to define the F-T interface type, it is recognized that many channels pass through a combination of land types before reaching the Bay.

Maps, photographs, and textual documents comprised the principal data types collected. (page 10: (a) Hesse 1861, courtesy of The Bancroft Library, UC Berkeley; (b) Rodgers 1856, courtesy of NOAA; (c) BANC MSS Land Case Files 87 ND, courtesy of The Bancroft Library, UC Berkeley; (d) Russell 1928-9, courtesy of Earth Sciences & Map Library, UC Berkeley); page 11: (e) USC&GS 1877, courtesy of NOAA)

e

Historical Fluvial-Tidal Interface Type

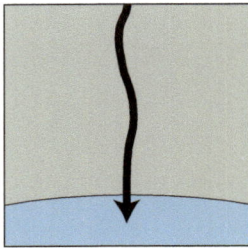

● **Connected to the Bay**

Channels entered directly into the Bay without passing through baylands (i.e., mudflats, tidal marshes, tidal-terrestrial transition zones).

Example: Hilarita Drainage (Marin County)

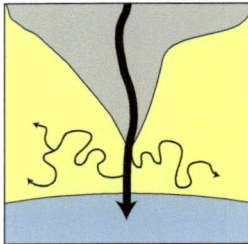

● **Connected to a tidal marsh channel**
◉ **with natural levee**

Channels reached tidal marshlands and merged into a tidal channel network.

Example: San Leandro Creek (Alameda County)
Example with levee: Guadalupe River (Santa Clara County)

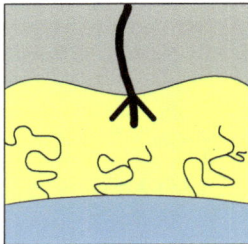

● **Drains onto a tidal marshland**
◉ **with natural levee**

Channels entered tidal marshlands and dissipated without connecting to a larger tidal channel network.

Example: Belmont Creek (San Mateo County)
Example with levee: San Lorenzo Creek (Alameda County)

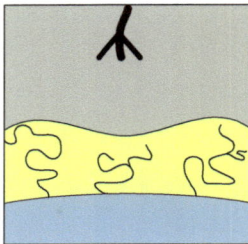

● **Disconnected on alluvial plain**
◉ **with natural levee**

Channels dissipated on alluvial plains or freshwater wetlands prior to reaching the baylands.

Example: Adobe Creek (Santa Clara County)
Example with levee: Stevens Creek (Santa Clara County)

Map of the historical fluvial-tidal interface types around San Francisco Bay. Markers are located either near the historical estuarine-terrestrial boundary (i.e., where the upland transitioned to tidal marsh) or at the historical channel terminus. An interactive version of this map can be found at *floodcontrol.sfei.org.*

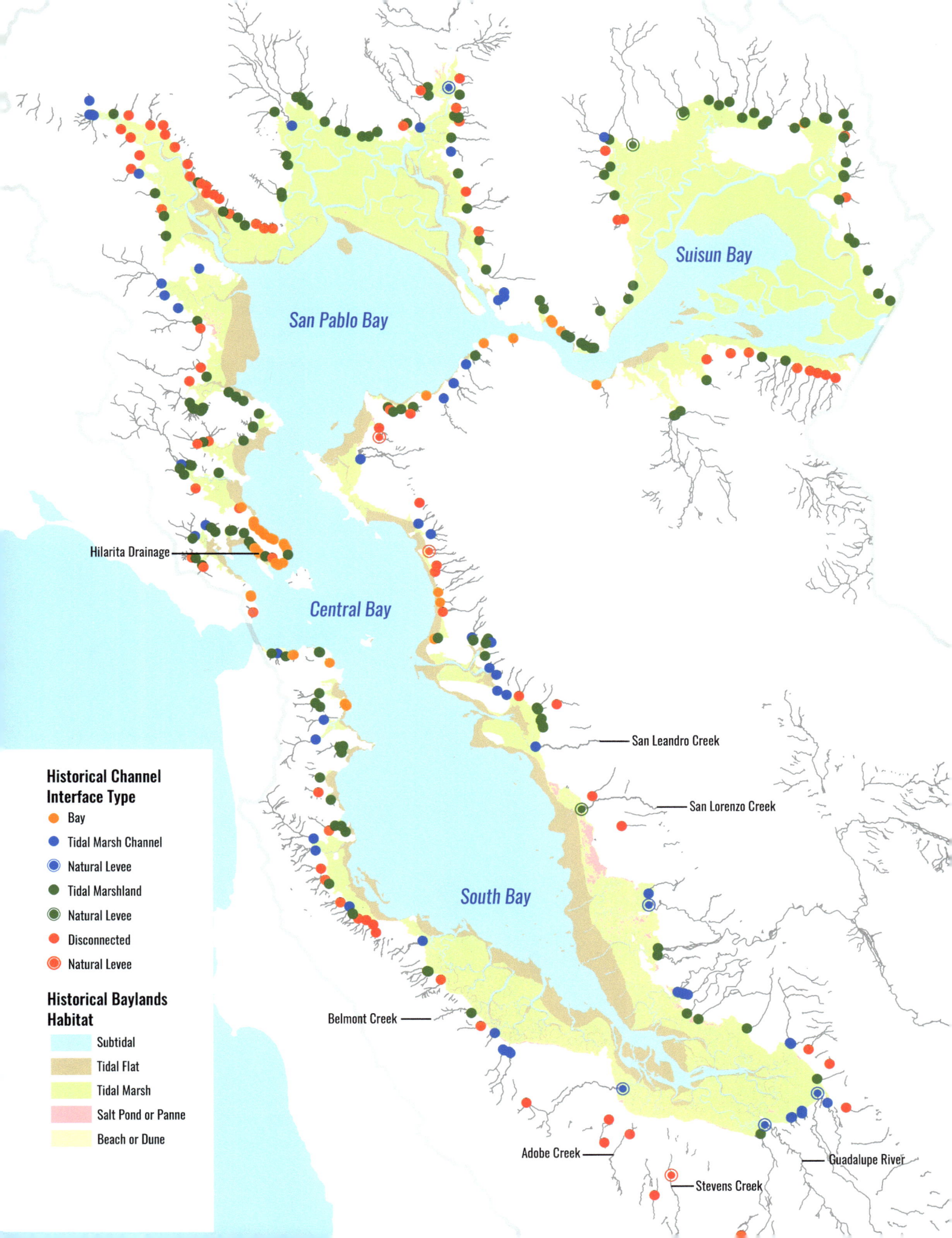

Historical Channel Interface Type
- 🟠 Bay
- 🔵 Tidal Marsh Channel
- ⊚ Natural Levee
- 🟢 Tidal Marshland
- ⊚ Natural Levee
- 🔴 Disconnected
- ⊚ Natural Levee

Historical Baylands Habitat
- Subtidal
- Tidal Flat
- Tidal Marsh
- Salt Pond or Panne
- Beach or Dune

Suisun Bay

San Pablo Bay

Central Bay

South Bay

Hilarita Drainage

San Leandro Creek

San Lorenzo Creek

Belmont Creek

Adobe Creek

Stevens Creek

Guadalupe River

Historical F-T Interface

Based on the historical mapping, we identified 353 historical creeks around San Francisco Bay. Of these, 47% drained directly onto tidal marshlands, 24% were disconnected from the tidal environment and dissipated on alluvial plains, 18% connected to a tidal channel network within tidal marshlands, and 11% entered directly into the Bay. Channels with natural levees were relatively rare and were most prevalent along channels that connected to a tidal channel network (9% of these channels), with 4% of disconnected channels and 2% of channels that drained directly onto tidal marshlands also having natural levees.

A range of physical factors contributed to the diversity of historical F-T interface types identified. The dominant factors controlling historical interface type appear to be the overall flow energy or stream power during storm events (inferred from the product of watershed area and overall landscape gradient at the F-T interface or channel terminus) and the overall watershed sediment supply per watershed area (inferred from watershed bedrock type and the presence of depositional features at the F-T interface). Here we describe key characteristics of each historical F-T interface type and discuss the relative magnitude of stream power and sediment supply.

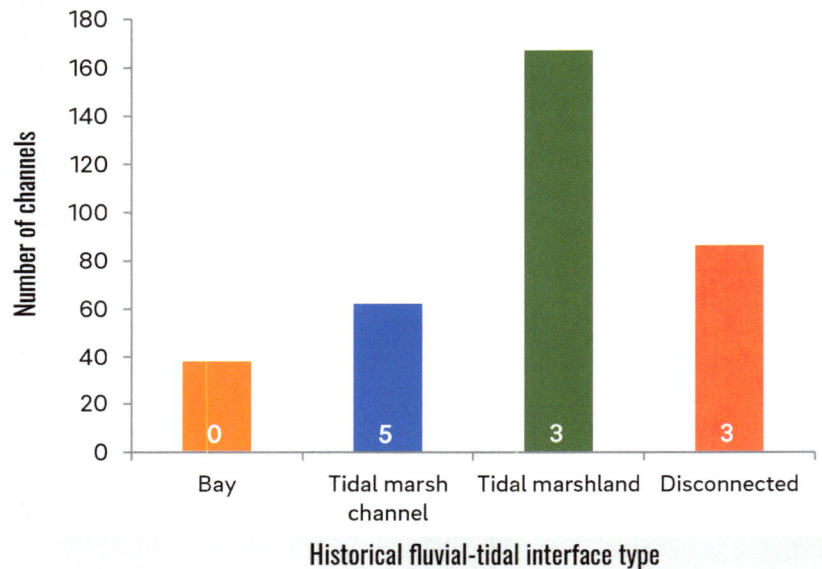

Summary of the number of San Francisco Bay channels within each historical fluvial-tidal interface type category. The numbers in white indicate how many channels had natural levees historically.

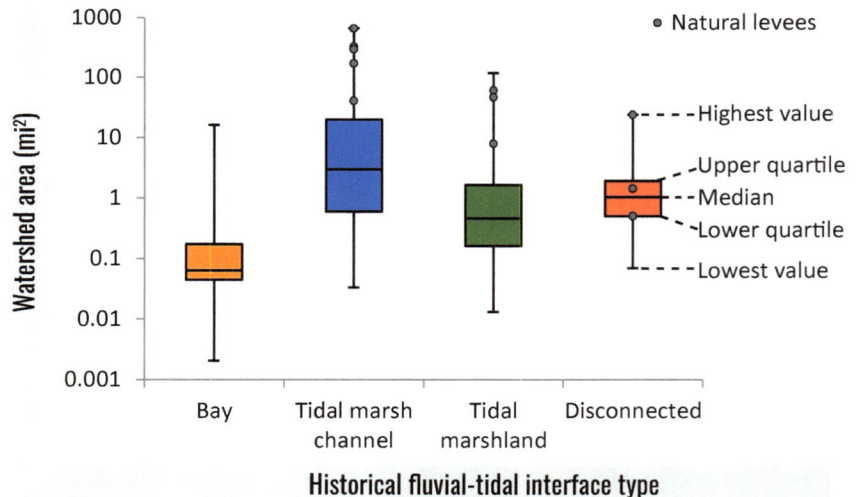

Box-and-whisker plot summarizing the watershed area statistics for each historical fluvial-tidal interface type.

CONNECTED TO THE BAY • This interface type was present in the Central Bay and Suisun Bay, with a high concentration along the Marin shoreline. It was typically associated with channels draining relatively small, steep watersheds (mean area = 0.7 mi^2, maximum area = 16 mi^2) that had relatively low stream power and relatively low sediment supply. In general, the small contributing watersheds did not have the sediment supply or the coastal morphology needed to sustain a marsh plain at their interface with the Bay or build natural levees.

CONNECTED TO A TIDAL MARSH CHANNEL • This interface type was historically present all around San Francisco Bay except in Suisun Bay. It was typically associated with channels draining relatively large watersheds (mean area = 35 mi^2, maximum area = 650 mi^2) that had a relatively high stream power and either a high or low sediment supply. In general, the watersheds that had considerable flow energy during storm events (i.e., the largest watersheds with a lot of storm flow) were able to maintain a discrete channel to and permanent connection with a tidal channel network, regardless of the watershed sediment supply. The largest watersheds had the highest sediment supply and therefore had natural levees at the F-T interface.

CONNECTED TO TIDAL MARSHLANDS • This interface type was historically present all around San Francisco Bay (with a high concentration in Suisun Bay) and was typically with relatively moderate sized watersheds (mean area = 3 mi^2, maximum area = 94 mi^2). In San Pablo Bay, Central Bay, and South Bay, this interface type was associated with watersheds that had relatively low stream power and low sediment supply. These channels typically drained watersheds with low channel gradients that caused sediment deposition at the marsh plain edge during storm events. In Suisun Bay, this interface type was associated with watersheds that had high stream power and either a high or low sediment supply. These characteristics are associated with channels elsewhere in the Bay that were connected to a tidal channel network, which is an interface type that was not present in Suisun Bay. It's possible that many of these Suisun Bay channels would have been connected to a tidal channel network if the brackish marshes around Suisun Bay were able to support extensive tidal channel networks like elsewhere in the Bay. In general, the largest watersheds had the highest sediment supply and therefore had natural levees at the F-T interface.

DISCONNECTED • This interface type was fairly evenly distributed around San Francisco Bay and typically associated with relatively moderate sized watersheds (mean area = 2 mi^2, maximum area = 24 mi^2). For some channels, the termination location at the bottom of the watershed was several miles from the Bay. Throughout most of the Bay, this interface type was associated with watersheds that had relatively low stream power and a relatively high sediment supply; in Suisun Bay, this interface type was associated with low stream power and both high and low sediment supply. During most storm events, many of these channels lacked the flow energy required to transport their high sediment load across alluvial plains, causing them to terminate in alluvial fans often miles from the Bay. Other channels lost stream power as they entered alluvial plains and lost their flow as it infiltrated into deep groundwater basins. Channels draining watersheds with steep topography and erosive geology had the sediment supply needed to build natural levees at the channel terminus.

Contemporary Fluvial-Tidal Interface Type

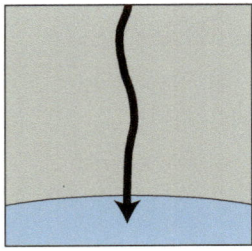

● **Connected to the Bay**

Channels entered directly into the Bay without passing through baylands (i.e., mudflats, tidal marshes, tidal-terrestrial transition zones).

Example: Hilarita Drainage (Marin County)

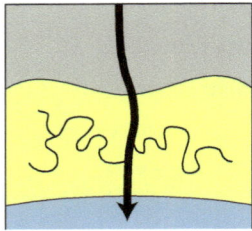

● **Connected to a tidal marsh channel**

Channels reach the baylands and merge into a tidal channel network.
Example: Petaluma River (Sonoma County)

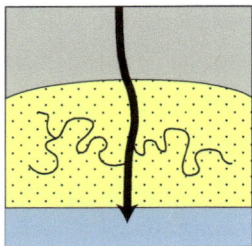

▶ **Connected to a tidal channel through diked baylands**

Channels enter areas where baylands have been diked (i.e., isolated from the tides by dikes or levees) and flow into a tidal channel.

Example: Novato Creek (Marin County)

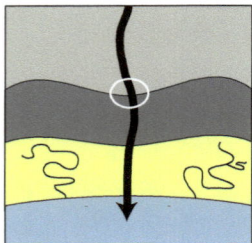

▶ **Connected to a tidal channel through bay fill**

Channels flow through bay fill (i.e., fine sediment placed on baylands to increase elevation and allow for development) before reaching the Bay.

Example: Wildcat Creek (Contra Costa County)

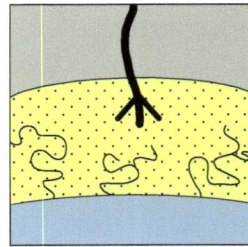

■ **Drains onto diked baylands**

Channels enter baylands that are now diked (e.g. salt ponds, managed marsh) but dissipate without connecting to a tidal channel.

Example: Willow Creek (Contra Costa County)

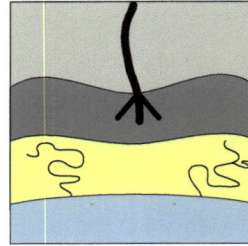

■ **Drains onto bay fill**

Channels enter baylands that are now covered in bay fill but dissipate without connecting to a tidal channel.

Example: Unnamed drainage to Scottsdale Pond (Novato Creek watershed, Marin County)

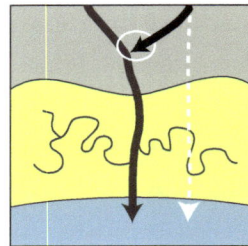

▶ **Channel has become a tributary channel**

Channels that historically reached the baylands but have been re-routed inland to flow into another channel.

Example: Laurel Creek (Solano County)

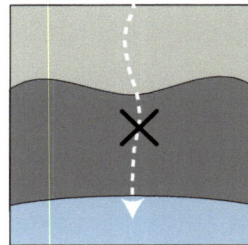

✕ **Channel no longer present on the landscape**

Channels have been routed into underground culverts or have been filled in completely.

Example: Yosemite Creek (San Francisco County)

Map of the contemporary fluvial-tidal interface types around San Francisco Bay. To allow for direct comparison with the historical interface map, markers are located at the historical interface locations. An interactive version of this map can be found at *floodcontrol.sfei.org*.

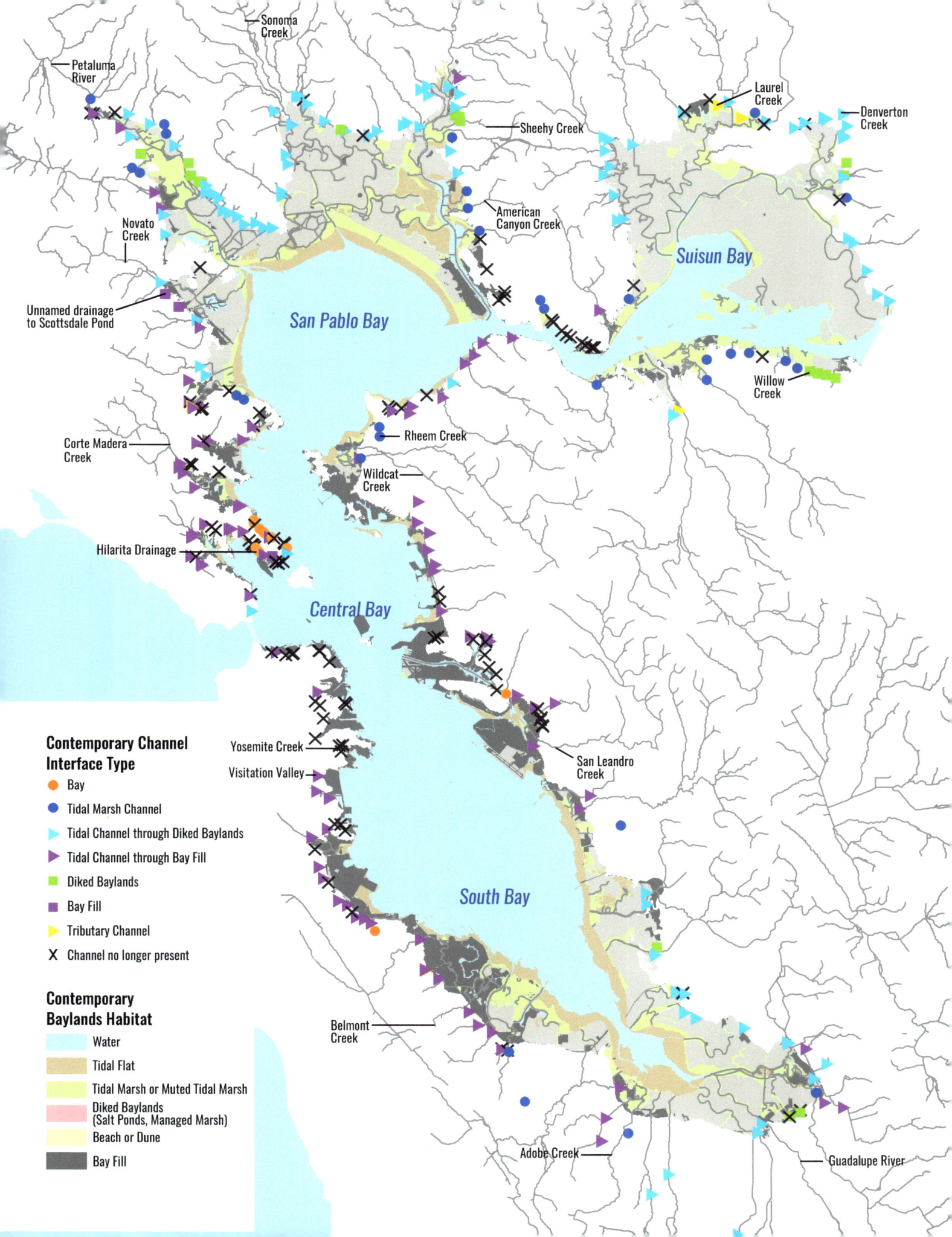

Sonoma
Creek

Petaluma
River

Laurel
Creek

Denverton
Creek

Sheehy Creek

Novato
Creek

American
Canyon Creek

Suisun Bay

San Pablo Bay

Unnamed drainage
to Scottsdale Pond

Willow
Creek

Corte Madera
Creek

Rheem Creek

Wildcat
Creek

Hilarita Drainage

Central Bay

Yosemite Creek

San Leandro
Creek

Visitation Valley

Contemporary Channel
Interface Type

- ● Bay
- ● Tidal Marsh Channel
- ▶ Tidal Channel through Diked Baylands
- ▶ Tidal Channel through Bay Fill
- ■ Diked Baylands
- ■ Bay Fill
- ▶ Tributary Channel
- ✕ Channel no longer present

South Bay

Contemporary
Baylands Habitat

- Water
- Tidal Flat
- Tidal Marsh or Muted Tidal Marsh
- Diked Baylands
  (Salt Ponds, Managed Marsh)
- Beach or Dune
- Bay Fill

Belmont
Creek

Adobe Creek

Guadalupe River

Contemporary F-T Interface

While some present day channels have the same F-T interface type as they did historically, the F-T interface type for the vast majority of channels has changed due to landscape alterations associated with flood control, agriculture, urban development, and diking of baylands for salt production and agricultural purposes. Of the 353 channels assessed, only 12% have an interface type that was present in the historical landscape (31 channels currently connect to a tidal channel and 11 channels currently connect directly to the Bay). The rest of the channels now flow into tidal channels adjacent to diked or filled historical marshes (51%), flow into diked baylands and bay fill (6%), were re-routed and are now tributaries to other channels (2%), or have been routed underground or filled in and no longer present (29%). Of the channels that remain on the landscape, 42% were historically disconnected from the Baylands and Bay except during extreme flood events.

Here we discuss the distribution of contemporary F-T interface types around the Bay and provide an overview of the change from historical to contemporary conditions associated with contemporary F-T interface type.

(top) Summary of the number of San Francisco Bay channels within each contemporary fluvial-tidal interface type category.

(bottom) Summary of the F-T interface conversion from historical to contemporary conditions. For example, of the 11 channels that currently connect to the Bay, nine historically connected to the Bay, one historically connected to a tidal channel, and one was historically disconnected.

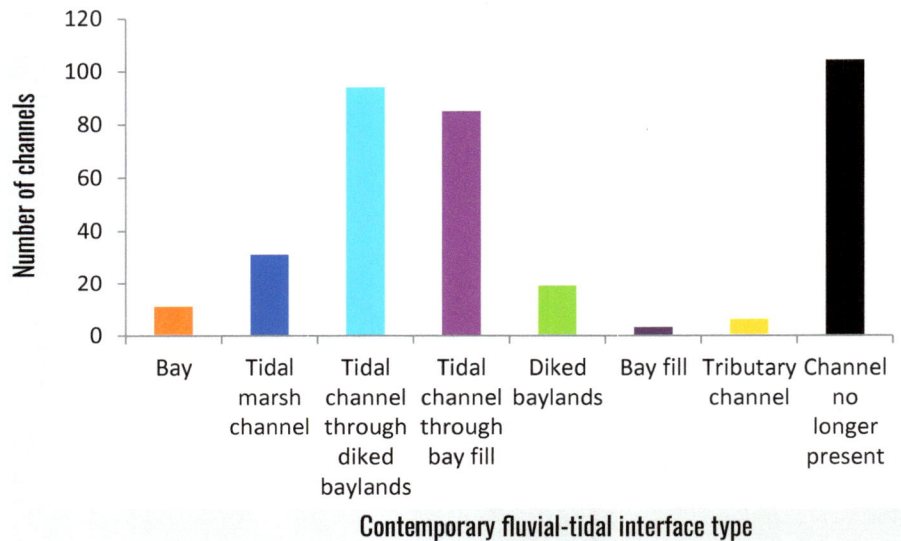

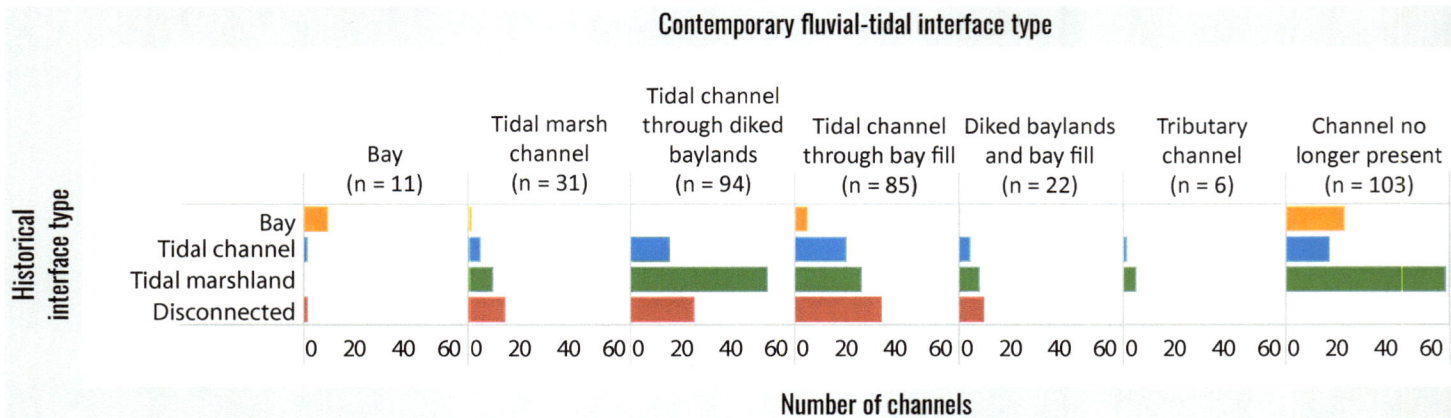

CONNECTED TO THE BAY • The vast majority of channels currently connected to the Bay had this same interface type historically. As in the historical landscape, most of the channels with this interface type today are at the bottom of relatively small watersheds draining the Marin headlands (e.g., Hilarita Drainage).

CONNECTED TO A TIDAL MARSH CHANNEL • A relatively small percentage of channels that currently connect to a tidal channel network had the same interface type historically (e.g., Petaluma River). Most were either historically disconnected (e.g., Rheem Creek) or were historically connected to baylands (e.g., American Canyon Creek). Channels that had this same interface type historically and that were historically disconnected are found all around the Bay. Channels that were historically connected to baylands now connect to tidal channels through relatively small marshes in San Pablo Bay and Suisun Bay.

CONNECTED TO A TIDAL CHANNEL THROUGH DIKED BAYLANDS • Most channels with this interface type historically emptied into baylands or were disconnected. These channels are primarily found adjacent to salt ponds in the South Bay (e.g., Guadalupe River) and in the Sonoma Baylands (e.g., Sonoma Creek), flowing into Petaluma marsh (e.g., Adobe Creek), San Pablo Bay (e.g., Novato Creek), and Suisun Pablo Bay (e.g., Denverton Creek).

CONNECTED TO A TIDAL CHANNEL THROUGH BAY FILL • Most channels with this interface type were historically disconnected or emptied into baylands. These channels are primarily found in the Central Bay in the highly urbanized areas around Redwood City (e.g., Belmont Creek), San Francisco (e.g., Visitation Valley Creek), Sausalito/Mill Valley/San Rafael (e.g., Corte Madera Creek), Pinole/Richmond (e.g., Wildcat Creek), and Berkeley/Oakland/Alameda (e.g., San Leandro Creek).

FLOWS INTO DIKED BAYLANDS AND BAY FILL • Channels that currently drain to diked baylands and bay fill were primarily disconnected or drained to baylands historically. Most channels are in diked agricultural areas at the mouths of the Petaluma River and Sonoma Creek in San Pablo Bay (e.g., Sheehy Creek), and in diked industrial areas around Suisun Bay (e.g., Willow Creek).

TRIBUTARY CHANNEL • The vast majority of channels that are now tributaries to larger channels historically drained to baylands (e.g., Laurel Creek); the rest historically connected directly to a tidal marsh channel. These channels are found all around the Bay and most were likely rerouted initially to increase available arable and pasture lands.

NO LONGER PRESENT • The majority of channels that are now in underground culverts or filled in once drained to baylands. The rest are split between channels that once drained straight to the Bay and channels that were connected directly to a tidal channel. Nearly all of these channels are associated with relatively small to moderate sized watershed (watershed area <2 mi^2). Routing channels underground and channel in-filling occurred all around the Bay, typically in highly urbanized areas around San Francisco (e.g., Yosemite Creek), Sausalito/Mill Valley/San Rafael, Vallejo/Benicia, Pinole/Richmond, and Berkeley/Oakland/Alameda.

SUMMARY AND SYNTHESIS

Major Findings

Over the past 200 years, channel reaches at the F-T interface around San Francisco Bay have been modified for land reclamation and flood control. The analysis presented here focused on identifying the past and present F-T interface types around the Bay and chronicling how interfaces have changed since the mid-19th century. The major findings from this analysis are as follows:

- **Historically, the dominant F-T interface types around the Bay were: 1) creeks that connected directly to the Bay; 2) creeks that connected to a tidal channel network; 3) creeks that drained onto tidal marshland; and 4) creeks that were unconnected to the tidal environment (except during large floods).** A small percentage of channels had enough sediment to build natural levees at the F-T interface. The vast majority of the channels included in the study once drained onto tidal marshlands, with these interface types being found all around the Bay and in a high density in Suisun Bay. Major drivers controlling the historical interface type include stream power during floods and watershed sediment supply. Overall, watersheds with relatively high stream power were typically associated with channels connected to a tidal channel network, watersheds with relatively low stream power and low sediment supply were typically associated with channels connected to the Bay or that drained on to tidal marshlands, and watersheds with relatively low stream power and high sediment supply were typically associated with disconnected channels.

- **Today, the F-T interfaces around the Bay look much different than they did in the past.** Although there are watershed channels that still connect directly to the Bay and to tidal marsh channel networks, most channels have been altered so that the connection is to a tidal channel that flows through diked baylands or bay fill or the channel has been routed underground or filled in completely at the historical F-T interface location. The channels that now connect to a tidal channel through diked baylands are typically associated with the salt ponds in the South Bay and diked areas in San Pablo Bay and Suisun Bay. The channels that now connect to a tidal channel through bay fill or that are no longer present on the landscape are typically in the highly urbanized shoreline areas around Redwood City, San Francisco, Sausalito/Mill Valley/San Rafael, Vallejo/Benicia, Pinole/Richmond, and Berkeley/Oakland/Alameda.

Management Implications

The historical and contemporary F-T interface classifications shown here are informative for understanding the magnitude of landscape change since the mid-19th century. They can also be used to help understand major drivers for contemporary sedimentation issues as well as potential opportunities for habitat restoration. Key management considerations derived from this study are as follows:

- **Major causes of sediment deposition and subsequent flood risk management issues.** There are several conclusions that can be drawn from this study regarding the inherent connections between F-T interface changes and sedimentation. First, the building of levees along channels that historically connected to a tidal channel network has often resulted in sedimentation issues that are linked to a decrease in tidal prism and associated increase in tidal sediment accumulation (e.g., Corte Madera Creek). Second, sedimentation issues in many historically disconnected channels are likely driven, at least in part, by the natural decrease in channel slope and associated stream power as the channels exit hills and flow onto alluvial plains (e.g., Adobe Creek in Santa Clara County). Third, sedimentation and flooding issues in places where channels have been filled in or routed underground are likely exacerbated by the loss of a path for sediment and water to exit to the Bay during flood events.

- **Opportunities for supporting habitat creation and restoration.** Many channels that historically connected to tidal marshlands or a tidal channel and are now constrained by levees are expected to still have the landscape setting (i.e., stream power) conducive for moving freshwater and sediment out to the Bay. These types of channels may provide the best opportunity for transporting freshwater and sediment needed to sustain created or restored baylands downstream over the long-term. In addition, many channels that had natural levees historically still have the high watershed sediment supply needed to build and maintain these features. Allowing that sediment to spread out onto adjacent tidal marsh plains during flood events would create topographic heterogeneity in the form of levees, splays and alluvial fans, and help build and maintain marshes. Developing habitat restoration concepts for these types of channels should ideally begin with an understanding of sediment supply to determine quickly if such restoration actions are viable. Estimates of contemporary watershed sediment supply for these types of channels are discussed in Chapter 3.

Black necked stilt in Corte Madera Creek, 2012. (Courtesy of Ketzirah Lesser and Art Drauglis, Creative Commons)

North Bay from above. (SFEI)

WATERSHED SEDIMENT YIELD AND SEDIMENT REMOVAL IN FLOOD CONTROL CHANNELS AT THE BAY INTERFACE

3

INTRODUCTION

In order to rethink the way flood control channels at the F-T interface around San Francisco Bay are designed and managed, a synthesis of information regarding sediment delivery to and removal from these channels is needed. In particular, developing new approaches for improving sediment delivery to baylands within the context of flood risk management requires a better understanding of the spatial and temporal dynamics of watershed sediment delivery to these channels and channel dredging for maintaining flood conveyance capacity, as well as an understanding of the cost associated with the current dredging practices. In this chapter, for the first time at a regional scale, we provide key information on the supply of sediment to flood control channels, sediment texture, and the amount of storage and removal over the past 50 or more years. This information provides new insights into a number of key management questions:

- How much sediment is delivered to each major flood control channel that drains to the Bay?

- How variable has the sediment load to these flood control channels been over the past several decades?

- How much, and at what frequency is sediment removed from each major flood control channel?

- Where along the major flood control channels is sediment being stored and removed?

- What is the grain size of removed sediment?

- How much does sediment removal cost?

METHODS

Flood Control Channel Selection

Information on sediment and channel characteristics was developed for 33 of the 353 channels described in Chapter 2. These 33 channels were selected because of interest by local flood control agencies, their known sedimentation issues, and their overall potential for supplying sediment that can help sustain the Bay's tidal marshes under a rising sea-level. The channels' "free flowing" watershed area (i.e., area downstream of any major dams) range in size from 1 to 370 mi^2 and represent 68% of the Bay Area local watershed area. Although the data set and analysis could be expanded in the future if quantitative data are collected from more watersheds and channels, the majority of the remaining 300+ channels drain relatively small watersheds and in many cases flow through underground pipes before discharging to the Bay. Management of the underground stormwater pipe infrastructure has a set of unique challenges and could be the subject of a future effort.

Sediment Delivery to Flood Control Channels

Inputs of coarse (>2 mm) and fine (<2 mm) watershed sediment to the 33 flood control channels were determined using local data collected by the United States Geological Survey (USGS), the Regional Monitoring Program (RMP), and local storm water agencies who hold National Pollutant Discharge Elimination System (NPDES) permits to discharge to the Bay. Measured suspended load and bedload data (together called total sediment load) were collated and synthesized to generate actual annual sediment load (tons/yr) for seven channels. The available data for these channels span a period between 1957 and 2013, with the number of years with suspended load and bedload data being very different for each channel (e.g., bedload data were collected

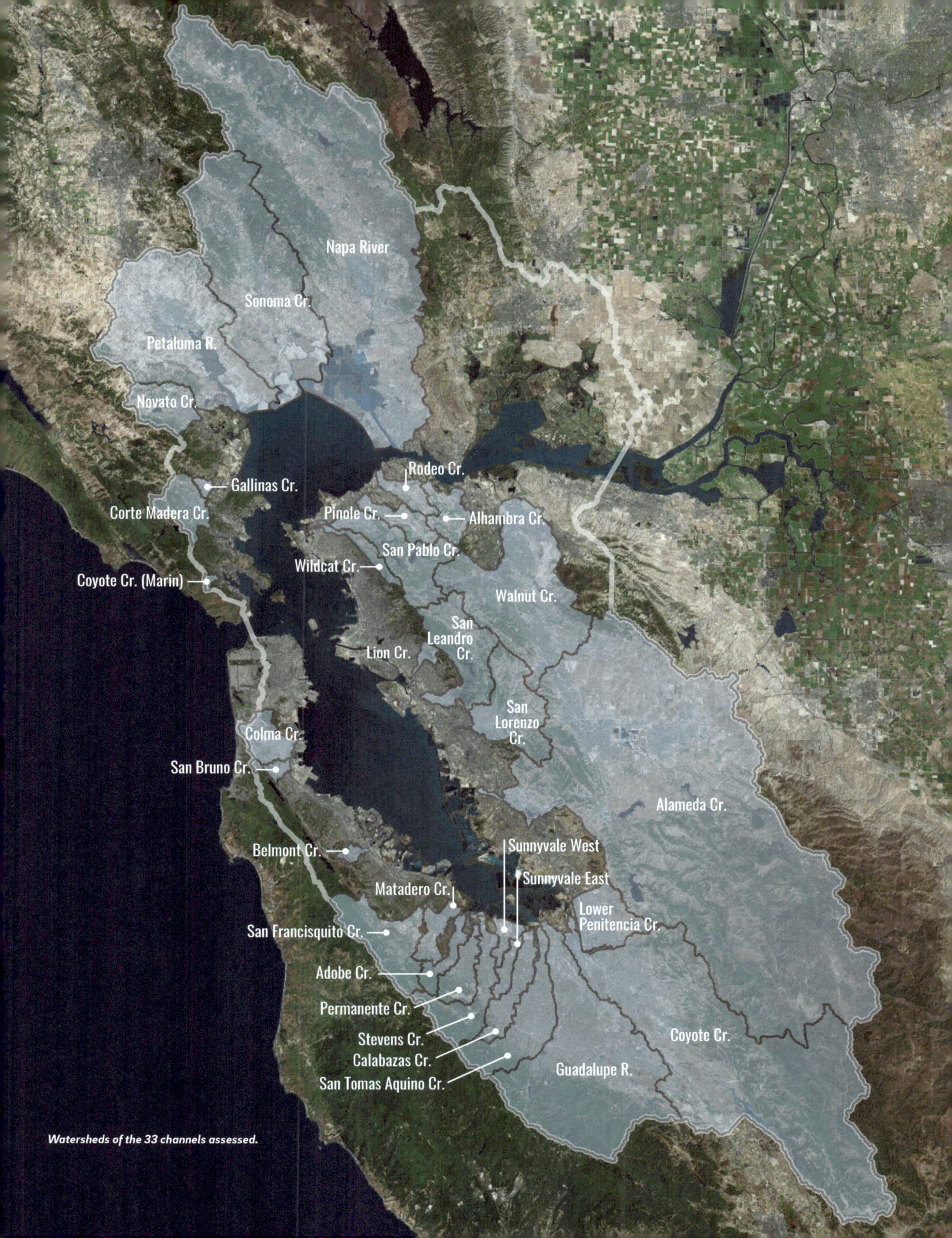

Napa River

Sonoma Cr.

Petaluma R.

Novato Cr.

Gallinas Cr.

Rodeo Cr.

Corte Madera Cr.

Pinole Cr.

Alhambra Cr.

San Pablo Cr.

Wildcat Cr.

Walnut Cr.

Coyote Cr. (Marin)

San
Leandro
Cr.

Lion Cr.

San
Lorenzo
Cr.

Colma Cr.

San Bruno Cr.

Alameda Cr.

Belmont Cr.

Sunnyvale West

Sunnyvale East

Matadero Cr.

Lower
Penitencia Cr.

San Francisquito Cr.

Adobe Cr.

Permanente Cr.

Stevens Cr.

Calabazas Cr.

San Tomas Aquino Cr.

Coyote Cr.

Guadalupe R.

Watersheds of the 33 channels assessed.

for one year on the Napa River and 13 years on Alameda Creek). For the remaining channels with no measured data, regional regression equations were used to estimate annual sediment loads from flow, watershed size, and degree of watershed urbanization (building upon methods described by McKee et al. [2013]). Due to differences in data availability and quality for the channels assessed, many simplifying assumptions were applied during interpretation, including assumptions of stasis (i.e., historical data are representative of today's conditions) and that sediment data collected during narrow flow ranges can be extrapolated to capture conditions during higher flows. Sediment load data for the channels was then normalized by watershed area to generate annual sediment yield (tons/mi^2/yr). Average annual sediment loads were then computed from the annual sediment load and yield.

For this regional synthesis of watershed sediment delivery, we focused on the time period between water year (WY) 2000 and WY 2013[1]. This period was chosen because it captures typical flow and sediment yield variability for Bay Area creeks (i.e., it is a period with several years above and below the long-term average annual precipitation recorded at San Francisco, CA [Golden Gate Weather Services 2016]) and is illustrative of the contemporary channel and watershed management regimes that affect sediment delivery dynamics. More information about the sediment load and yield data can be found in Appendix A (Table 1) and the sediment database and associated metadata.

Sediment Deposition, Removal, Cost, and Grain Size

Data on sediment removal quantities, location, dates, costs, sediment grain size, and in-channel deposition were obtained by reviewing reports prepared by flood control, city, and county agencies and their consultants and by conducting interviews with agency staff (Appendix A, Table 2). Data from various sources was combined so that a chronological history of sediment removal dates, volumes, costs (if recorded), grain size (if available), and re-use or disposal information (if available) for each channel could be created. Reported deposition estimates were from comparisons between as-built channel dimensions and recent channel surveys. These data were then quality checked.

For each channel, the head of tide location (i.e., the approximate inland extent of tidal inundation during MHHW) was defined so that sediment data could be classified as occurring in the fluvial or the tidal portion of the channel. Head of tide was defined in most channels using agency knowledge along with physical and biological indicators derived from aerial photograph inspection. However, in some channels where no local knowledge existed, the NOAA sea-level rise viewer (https://coast.noaa.gov/slr/) was utilized. This viewer allows the user to access a national data set on the position of mean higher high water (MHHW), a reasonable estimate of the mean position of head of tide assuming limited influence of frictional roughness and no tide control structures (e.g., gates and weirs). For the few sediment removal events that spanned the head of tide, we divided the removed volume into tidal and fluvial based upon the proportion of channel length where sediment was removed.

---

[1] A water year extends from October through the following September and is identified with the calendar year in which it ends (e.g., water year 2016 extends from October 1, 2015 to September 30, 2016).

Sediment deposition data, including location, date, volume, and grain size were also requested and gathered. Net deposition is defined as the total amount of sediment currently "stored" in each channel, based upon comparisons between as-built dimensions and recent channel surveys. However, very few channels had enough (or any) data for this calculation, or only had data for a portion of the entire flood control channel length. Those that did have data often were recording volumes of sediment that accumulate in between removal events, which were not consistent, and thus were not recording net deposition. Given these challenges, we note that although deposition data are recorded in the database, they will primarily be useful for the temporal analysis of a single channel. We caution against comparison between channels.

Considering all channels together, the compiled information spanned a period from 1958 to 2013 with unique start date for each individual channel designated by when a channel was first built or when the first recorded

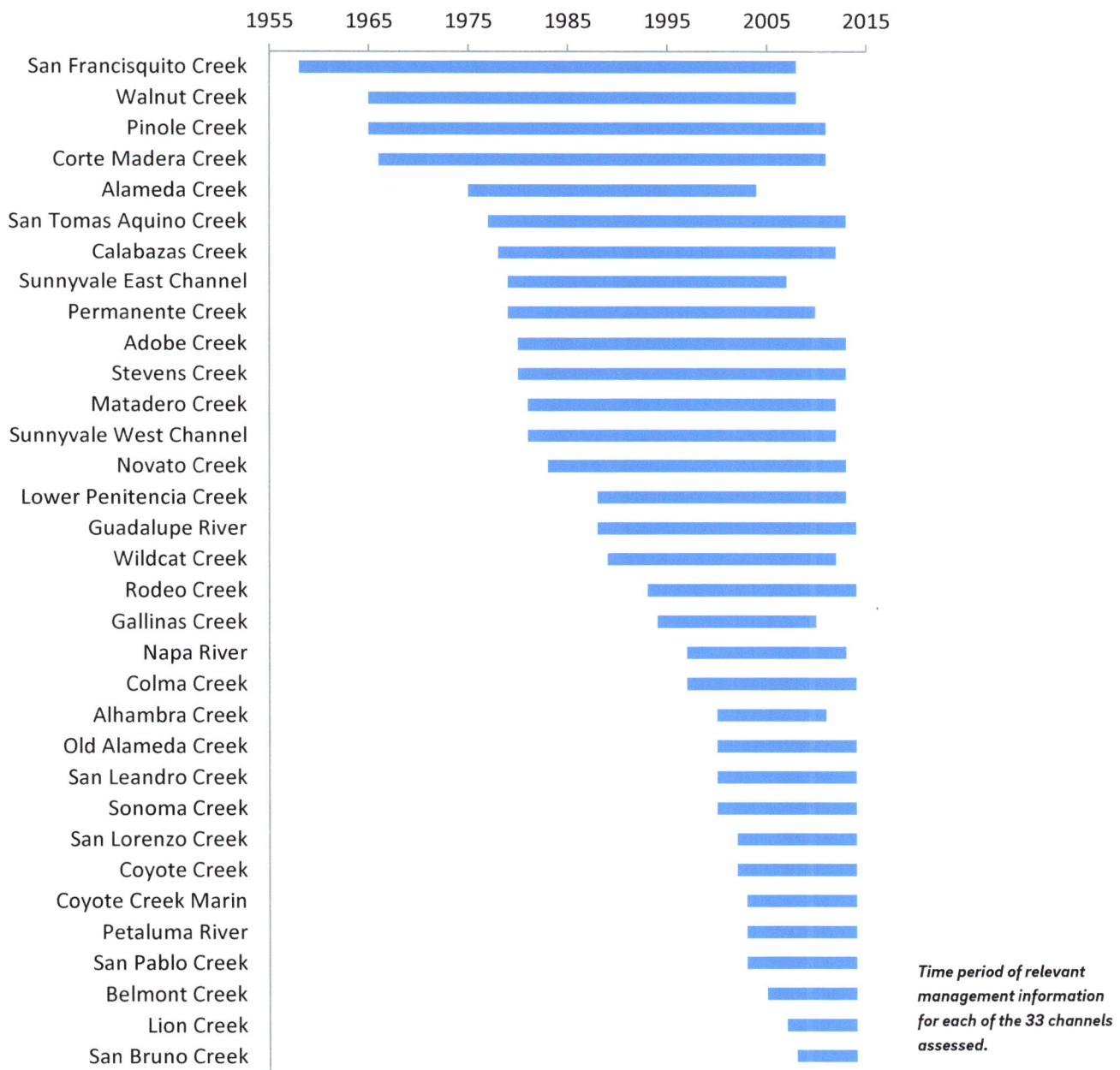

Time period of relevant management information for each of the 33 channels assessed.

sediment deposition or removal began. For each channel, the number of years with data ranged widely: San Bruno Creek had the shortest record (just six years), whereas the four creeks with the most data (San Francisquito, Walnut, Pinole, and Corte Madera Creeks) have been observed for over 45 years each. Although there was variability in data quality across the data sets, we generally have low confidence for sediment storage, medium confidence for the costs of sediment removal, and high confidence for sediment removal volume information. However, very few measurements of grain size exist for deposited or removed sediment.

RESULTS

Interannual Variation in Sediment Load Delivery

Annual suspended sediment load and bedload were measured or estimated from WYs 1957 to 2013 for all 33 channels. Bay Area watersheds experience considerable interannual climate variation, making it quite common for watersheds to exhibit annual sediment loads that vary by 10-fold (e.g., Colma Creek at South San Francisco) to well over 100-fold (i.e., Napa River at Napa) between successive years. Although geological variation also influences erodibility, in general, interannual sediment load variability increases with interannual flow variability, watershed size (more weakly), and decreases with amount of impervious surface. For example, based on data for watersheds with at least three years of both suspended and bedload data, it is evident that a large amount of the interannual sediment load variability can be estimated from measurements of peak flow variability. It is interesting to note that bedload is much more variable than suspended load, an observation that makes sense given the stochastic nature of sediment supply events (e.g. landslides and debris flows), and that bedload transport does not occur below a threshold of transport initiation.

Within urbanized watersheds (e.g., Colma Creek) we see low flow variability due to consistent runoff from impervious surfaces. Here, the erosion sources are minimized

Suspended sediment load variability and bedload variability vs. peak flow variability for several Bay Area creeks.

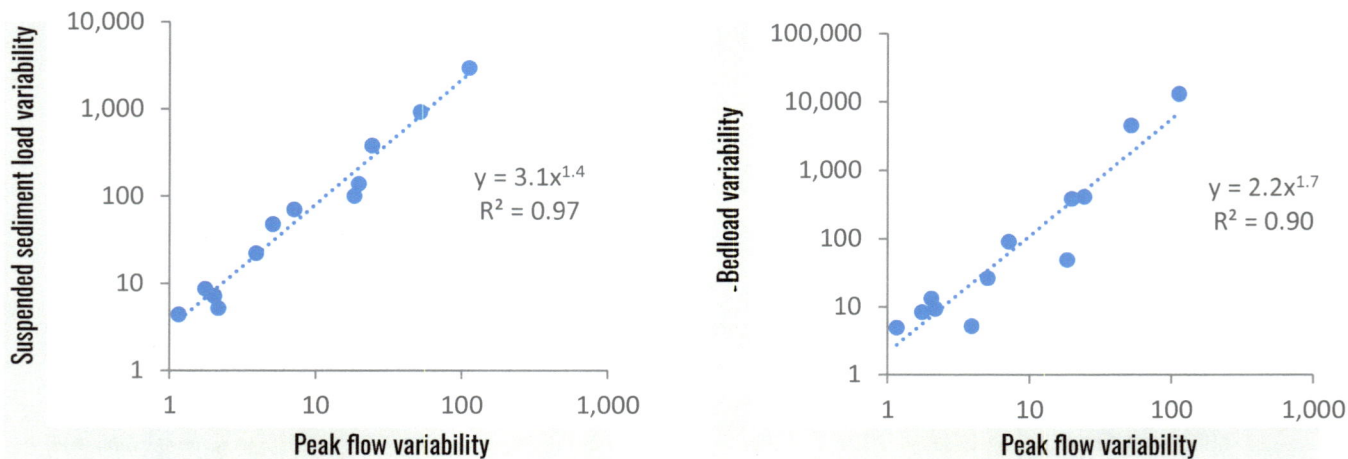

Left plot: $y = 3.1x^{1.4}$, $R^2 = 0.97$. X-axis: Peak flow variability. Y-axis: Suspended sediment load variability.

Right plot: $y = 2.2x^{1.7}$, $R^2 = 0.90$. X-axis: Peak flow variability. Y-axis: -Bedload variability.

due to an impervious built-out landscape, as well as the considerable efforts to stabilize slopes and control creek dimensions, thereby typically leading to relatively low inter-annual sediment load variability. Conversely, large and mostly non-urbanized watersheds (e.g., Napa River) exhibit both extreme flow variability and extreme spatial and temporal variability in sediment erosion that includes landslides, debris flows, and channel erosion sources. For example, the drought year of 1977 produced a peak flow of just 54 cubic feet per second (cfs) and an estimated sediment load of just 2 tons in the Napa River whereas the extremely wet year of 1986 produced a peak flow of 37,100 cfs and an estimated sediment load of around 1,000,000 tons. Comparing all 33 watersheds in this study (and based on estimated variability over a 40 year averaging period), we estimate the median interannual variation to be approximately 3,000-fold between the lowest and highest sediment producing years in a given watershed. This extreme variability will undoubtedly influence the potential use of fluvial sediment supply for any wetland restoration and make the redesign of the fluvial-tidal interface difficult.

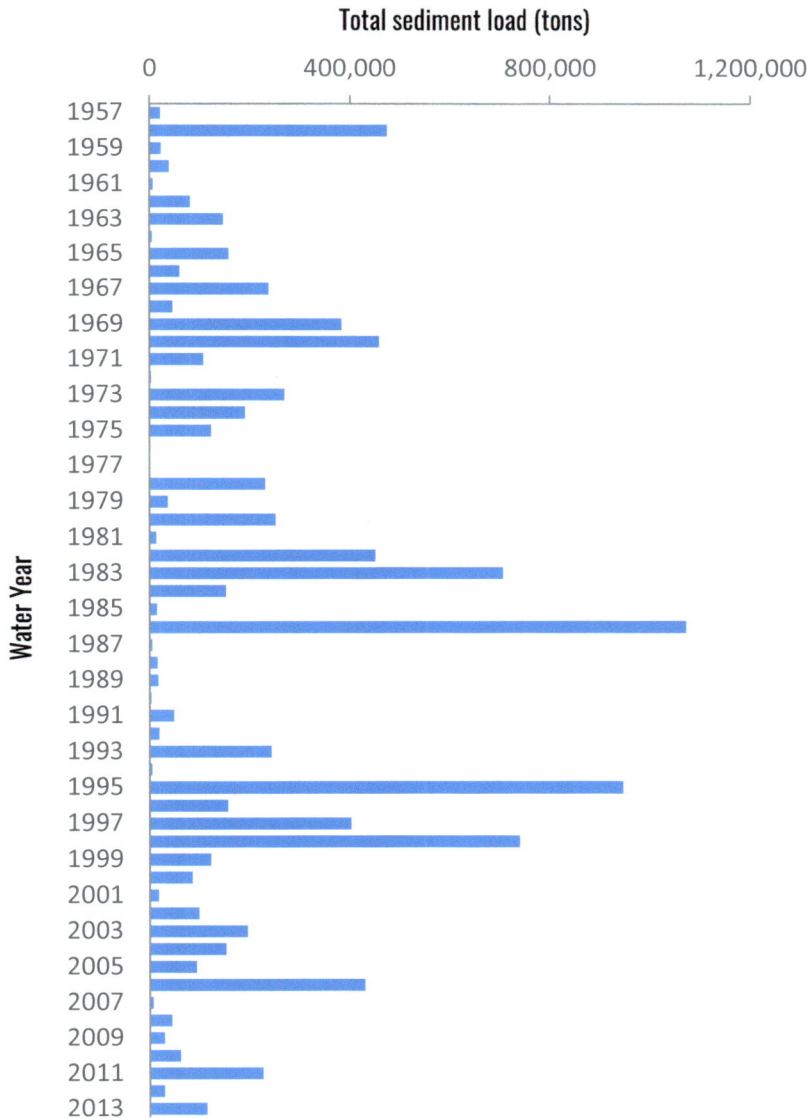

Napa River total annual sediment load estimates (1957-2013).

Average Annual Sediment Inputs to Flood Control Channels

A number of factors influence the magnitude of the total sediment loads coming into our flood control channels including watershed area geology, land management and use, discharge, and discharge variability from year to year. Our analysis shows that the lowest average annual sediment loads are being delivered to the Bay from a number of smaller watersheds that each delivered <4,000 tons total for the period 2000-2013 (Sunnyvale West Channel, Sunnyvale East Channel, San Leandro Creek, Gallinas Creek, Coyote Creek Marin, Belmont Creek, Lion Creek, Novato Creek, and San Bruno Creek). Within this group, several watersheds (e.g. San Leandro Creek) have particularly low average annual sediment loads presumably due to the reduction of effective watershed area associated with reservoir trapping. On the other extreme, a number of larger watersheds are delivering in excess of 45,000 tons on average per year (Sonoma Creek, Walnut Creek, Napa River, Alameda Creek, and Petaluma River). In the future, the small, highly urbanized watersheds and those with

Average annual watershed sediment loads at head of tide (2000-2013).

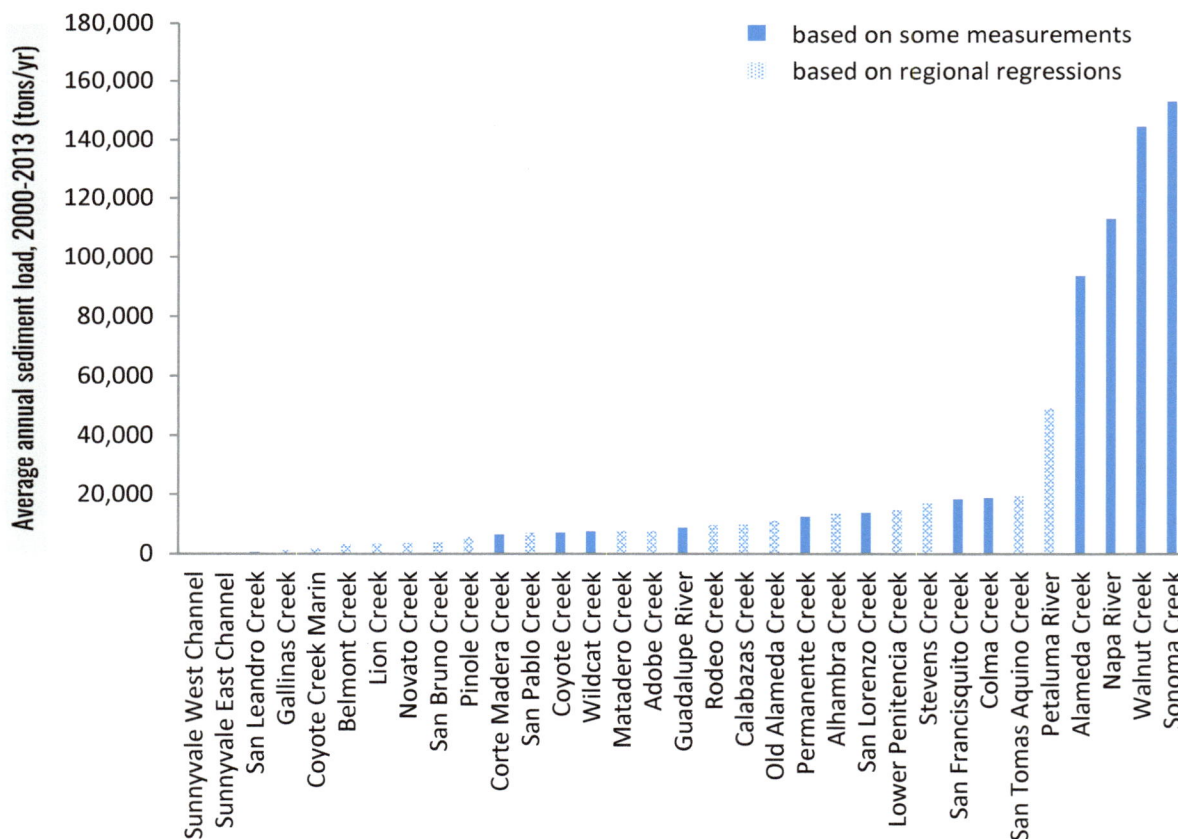

30

significant reservoirs will likely continue to deliver relatively low sediment loads, whereas the larger, less urbanized watersheds may experience greater variability, including potentially larger total sediment loads.

When normalized by contributing watershed area upstream of head of tide and below any major reservoirs, estimates of average annual yields for the period 2000-2013 for the 33 channels range between 50 and 1,660 tons/mi^2/yr (a 33-fold variation) in relation to watershed areas that range from 1.2 – 370 mi^2 (about a 300-fold variation). This variation in yield is indicative of variation in geology, climate, and land management in the Bay Area and provides a large challenge for the design of flood control channels in relation to sediment trapping and transmission. For example, the channel management of the tidal portion of the Guadalupe River (watershed area = 97 mi^2), which has a total sediment yield of 90 tons/mi^2/yr, will be inherently different than that of Sonoma Creek (watershed area = 92 mi^2), which has a total sediment yield of 1,660 tons/mi^2/yr.

Average annual watershed sediment yields at head of tide (2000-2013).

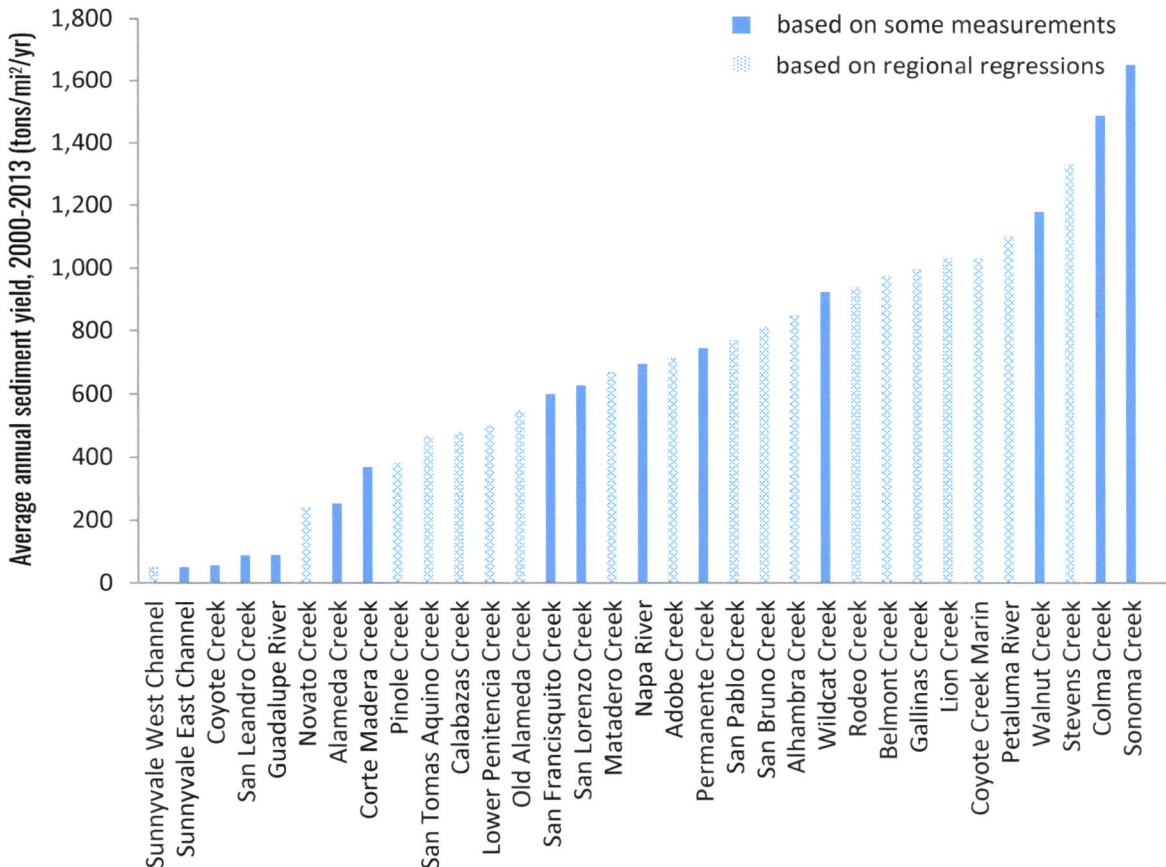

Sediment Deposition in Flood Control Channels

Information on sediment deposition in flood control channels is limited to just 14 channels: Alameda Creek, Alhambra Creek, Colma Creek, Corte Madera Creek, Coyote Creek Marin, Lion Creek, Gallinas Creek, Napa River, Petaluma River, Pinole Creek, Rodeo Creek, San Francisquito Creek, Walnut Creek, and Wildcat Creek. The total amount of recorded sediment deposition ranges from 1.9 million cubic yards (CY) in Alameda Creek Flood Control Channel to 1,145 CY in Alhambra Creek (for all channels we assumed a bulk density of 1.25 tons/CY for finer grained sediment and 1.40 tons/CY for coarser grained sediment). For many channels in the study (e.g., Alameda Creek), some or all of the sediment recorded as deposition has already been removed. For the other channels, there is either a general lack of as-built cross sections and longitudinal profiles or there have been no systematic repeat surveys to determine the volume of sediment deposition. This lack of data pertains to both sufficiently-sized channels that have experienced deposition and equilibrated to the new condition without new flooding, and to those that have experienced deposition and have lost significant channel flood capacity. As sea-level continues to rise, the F-T transition will move upstream, and may have a significant effect on the deposition of sediment in both the fluvial and tidal reaches.

Volumes of Sediment Removed from Flood Control Channels

Over the past several decades, approximately one-quarter of the 33 channels assessed have had sediment removed from only the tidal reach below head of tide, approximately one-quarter have had sediment removed from only the fluvial reach just upstream of head of tide, and approximately one-half have had sediment removed from both reaches. Since 1973, a total of 5.8 million CY of sediment was removed from 30 of the 33 channels assessed (three channels had no data), with 63% of that sediment removed from tidal reaches. From 2000 to 2013, a period more representative of the contemporary policy and management paradigm, 1.7 million CY of sediment has been removed, 72% of which was removed from tidal reaches. Of this total, most of this sediment came from 9 channels that had total removal volumes greater than 50,000 CY (Alameda Creek, Walnut Creek, Petaluma River, Gallinas Creek, Novato Creek, San Tomas Aquino Creek, Napa River, Old Alameda Creek, and Sunnyvale East Channel). Although some sediment has been used for bayland restoration (e.g., salt marsh restoration or filling of borrow ditches) and other habitat restoration purposes, we find that most sediment (>60%) is currently being disposed of as a waste product.

In addition to the variability in total volume and location of sediment removal, the frequency and management of sediment removal varies from channel to channel. Some channels, like Alameda Creek, have sediment removal in response to large flood events, when sediment is deposited in the channel and flood conveyance capacity is lost. The channels with a very irregular schedule of removal also tend to have irregular total volumes removed. Other channels, like Novato Creek, have a prescribed removal schedule where roughly the same volume is removed every few years. Still other channels, like those in Santa Clara County, have sediment removal as part of a permitted management regime and relatively small sediment volumes removed on a regular basis (typically every other year). From 1973 to 2013, approximately one-third of the sediment

Percent of sediment removed (y-axis)

Legend:
- Removed from tidal reach
- Removed from fluvial reach

Bar chart categories (x-axis): San Leandro Creek (No Data), San Pablo Creek (No Data), Sonoma Creek (No Data), Colma Creek, Rodeo Creek, Sunnyvale West, San Francisquito Creek, Las Gallinas Creek, Petaluma River, Napa River, Sunnyvale East, Corte Madera Creek, Alhambra Creek, Coyote Creek Marin, Novato Creek, Old Alameda Creek, San Tomas Aquino Creek, Adobe Creek, Walnut Creek, Guadalupe River, Coyote Creek, Lower Penitencia Creek, Stevens Creek, Matadero Creek, Alameda Creek, Calabazas Creek, Pinole Creek, San Bruno Creek, Belmont Creek, San Lorenzo Creek, Lion Creek, Permanente Creek, Wildcat Creek

Pie chart: Responsive to storms, 20%; 1 to 5 years, 35%; 5 to 15 years, 45%

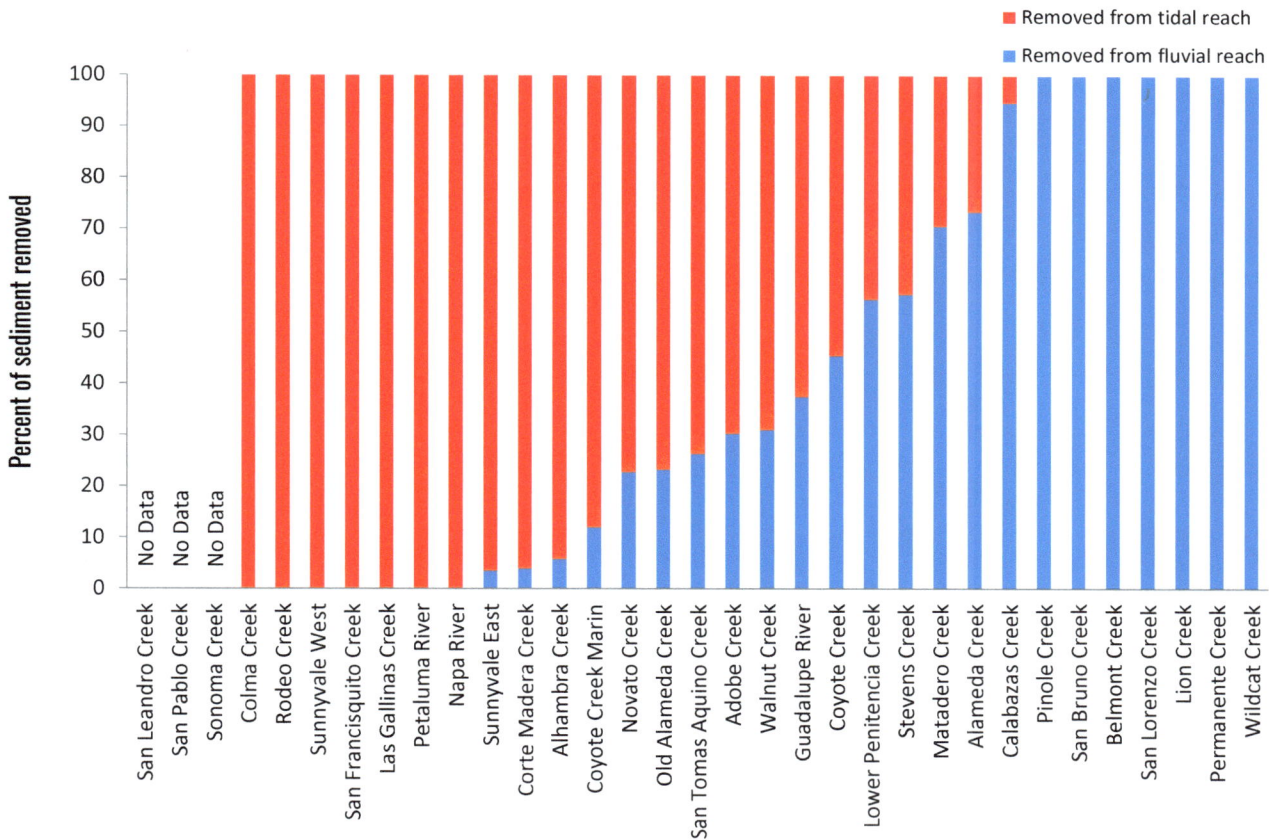

(Top) Percent of sediment removed from the tidal reach downstream of head of tide and the fluvial reach upstream of head of tide (1973-2013).

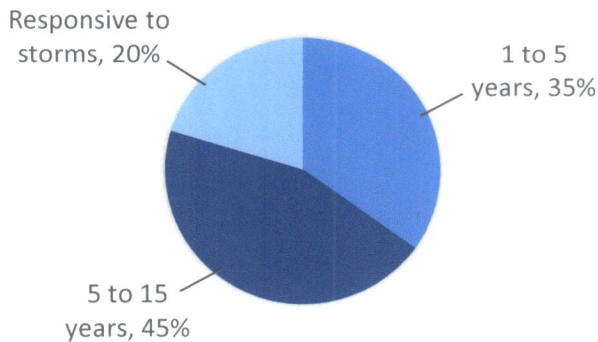

(Bottom) Percent of the total sediment removed by removal frequency (1973-2013).

came from channels with a somewhat regular sediment removal frequency of one to five years and almost half came from channels with a somewhat regular sediment removal frequency of five to 15 years. During this time, almost all of the sediment removed from channels that were dredged in response to large storms occurred, on average, at a frequency less than every five years (with most of that sediment coming from Alameda Creek). Therefore, between 1973 and 2013, more than half of the sediment came from channels that were dredged, on average, at least once every five years. However, projects that change the channel morphology (e.g., creation of multi-stage channels) may increase sediment transport capacity, and thus reduce the need for sediment removal in response to large storm events for some of these channels.

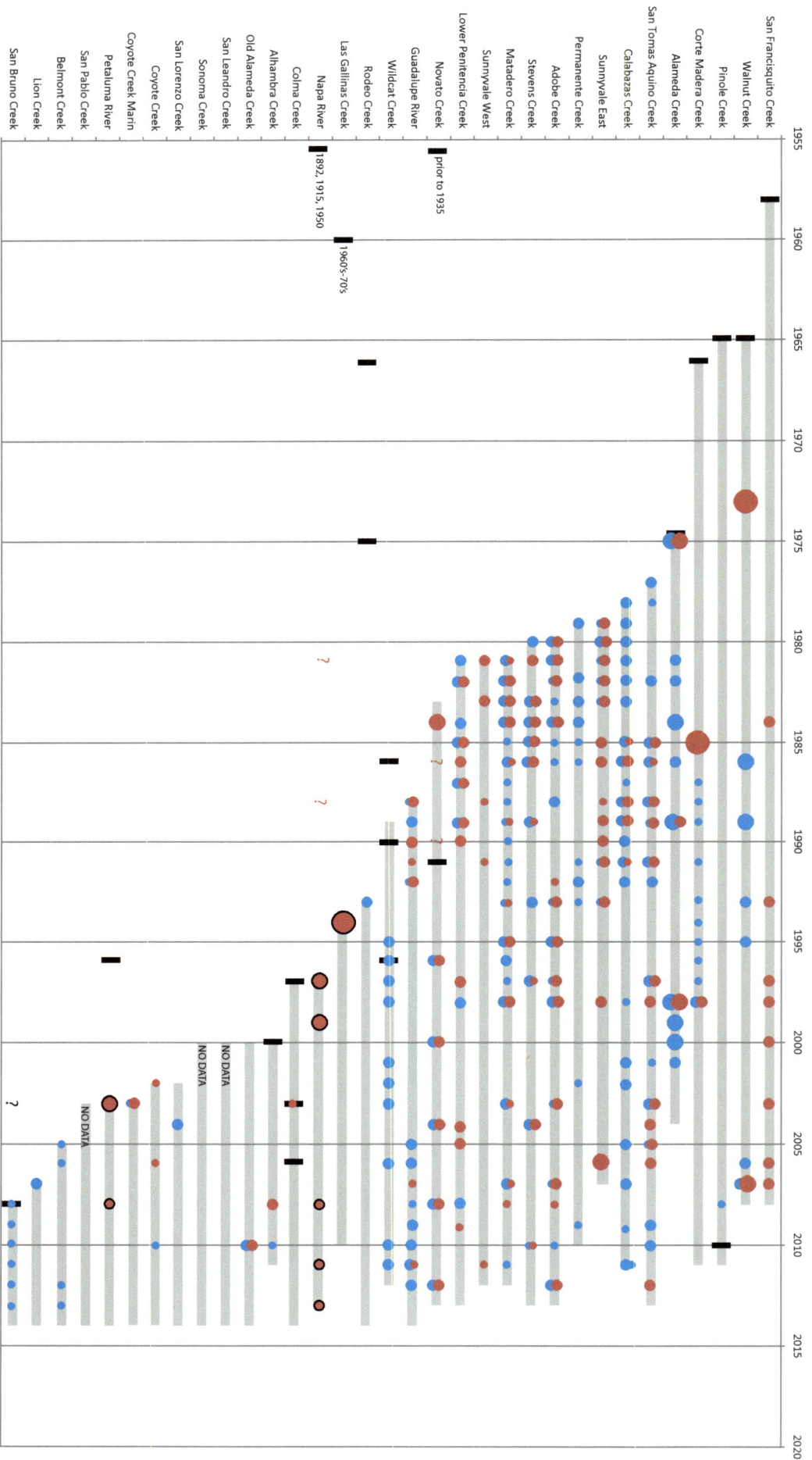

Sediment removal history (1957–2013), showing removal magnitude and location (fluvial or tidal) for each removal event.

Budget Time Period

- Channel Built/Significant Alteration
- ? Known Removal, Unknown Amount

Removal from Tidal Portion
Removal from Fluvial Portion
Tidal Removal for Navigation

- <1,000 CY
- 1,000–50,000 CY
- 50,000–200,000 CY
- >200,000 CY

San Francisquito Creek
Walnut Creek
Pinole Creek
Corte Madera Creek
Alameda Creek
San Tomas Aquino Creek
Calabazas Creek
Sunnyvale East
Permanente Creek
Adobe Creek
Stevens Creek
Matadero Creek
Sunnyvale West
Lower Penitencia Creek
Novato Creek
Guadalupe River
Wildcat Creek
Rodeo Creek
Las Gallinas Creek
Napa River
Colma Creek
Alhambra Creek
Old Alameda Creek
San Leandro Creek
Sonoma Creek
San Lorenzo Creek
Coyote Creek
Coyote Creek Marin
Petaluma River
San Pablo Creek
Belmont Creek
Lion Creek
San Bruno Creek

1892, 1915, 1950
1960's–70's
prior to 1935

NO DATA

34

Grain Size of Fluvial Sediment Loads

Sediment grain size not only influences the propensity of sediment to be trapped in flood control channels but also influences the quality of the habitat within the channel and its re-use potential for baylands restoration. A channel's ability to transport or deposit sediment of different grain sizes varies along the channel's longitudinal profile. The channel gradient within the fluvial reach decreases from the steep headwaters downstream to the F-T transition, or head of tide location. As gradient decreases, the stream power also decreases, causing coarser sediment to deposit within the lower fluvial reaches while finer sediment is transported further downstream. Within the tidal reach, the channel gradient is very low and tidal prism controls channel geometry and overall fine sediment deposition dynamics in the channel and on adjacent benches between the outer channel banks and levees.

Grain size measurements have been made on watershed sediment supplied to just six of the 33 channels assessed (San Lorenzo Creek, Alameda Creek, Wildcat Creek, Napa River, Guadalupe River, and Corte Madera Creek). In general, data have been collected only during smaller storms when larger grain sizes may not have been very mobile and therefore not well represented in the data. Sediment data for a few locations (Cull Creek, San Lorenzo Creek, and Alameda Creek) show general downward trends in the ratio of suspended sediment to bedload in relation to peak flow, suggesting that coarse sediment becomes supply-limited as flows increase, although no data has been collected during very high flows in the Bay Area. Data collected on Alameda Creek at the Niles gauge shows that 60% of the load passing into the flood control channel is silt and clay (<0.0625 mm in diameter) and 8% is gravel and larger (>2mm).

Information on the grain size of sediment deposited in or removed from flood control channels is similarly sparse and only available in eleven channels (Alameda Creek, Alhambra Creek, Colma Creek, Corte Madera Creek, Novato Creek, Petaluma River, Pinole Creek, San Bruno Creek, San Francisquito Creek, Walnut Creek, and Wildcat Creek). Most of the data is qualitative (i.e., has an indication of fine or coarse) and is generalized for the fluvial and the tidal reaches, but some creeks, like Alameda Creek, do have more quantitative data. Sediment samples taken in Alameda Creek downstream of Niles Canyon show that, on average, 40% of bed sediment is gravel and larger (>2 mm). Sediment caught in the channel becomes finer moving downstream, ranging from ~70% gravel and larger in the upper reaches to <20% gravel and larger in the downstream reach near head of tide. For this and other channels that trap coarse sediment, there could be opportunities to change the channel design and management to move more of this sediment downstream so that it builds depositional fans at the Bay margin and helps maintain beaches along marsh edges. However, sea-level rise will cause the head of tide location to move inland, which will result in channel slope reduction in the downstream tidal reaches and present challenges for coarse sediment transport. Additionally, the effects of future climate change make the effects on grain size difficult to predict; there may be changes in the processes sourcing sediment (e.g., landslides, channel incision) or changes in the discharge, and thus ability to transport different grain sizes.

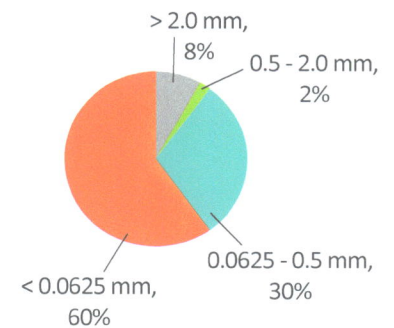

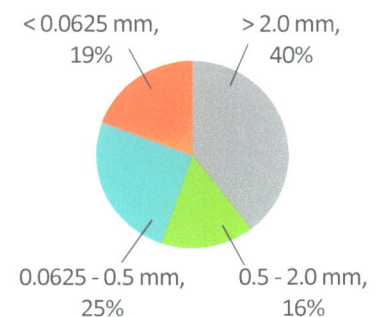

(Top) Size distribution of total sediment load (suspended load and bedload) recorded at the Alameda Creek at Niles Canyon gage (USGS 11179000) (1965-2013).

(Bottom) Bed sediment size distribution for Alameda Creek downstream of the Niles Canyon gage (USGS 11179000) (2009). Data collected by SFEI.

Cost of Sediment Removal from Flood Control Channels

Channel maintenance for flood conveyance is costly and requires permits that are increasingly difficult to obtain mainly due to concerns over habitat disturbance without compensatory activities. The 5.8 million CY of sediment removed since 1973 cost $111 million (not adjusted for inflation) or about $2.8M per year (excluding Sonoma, San Leandro, and San Pablo Creeks). Napa River, Walnut Creek, and Alameda Creek together account for 63% of those total costs. In an attempt to make some sense of the data and compare channels, the volumetric sediment removal totals were normalized to the area of the flood control channels. Costs range from $1,225 to $5,459,000 per square mile of channel dredged per year, with an average of $705,000. Several channels (e.g., Lower Penitencia and Lion Creeks) appear to have the highest cost of operation in relation to channel area. Lower Penitencia Creek has had 14 removal events from 1981 to 2013 whereas Lion Creek only had a one-time removal that cost over $600,000. Sediment from Lion Creek went to landfill whereas most recently, regular sediment removal from Penitencia Creek has been used for wetland restoration in the South Bay Salt Ponds.

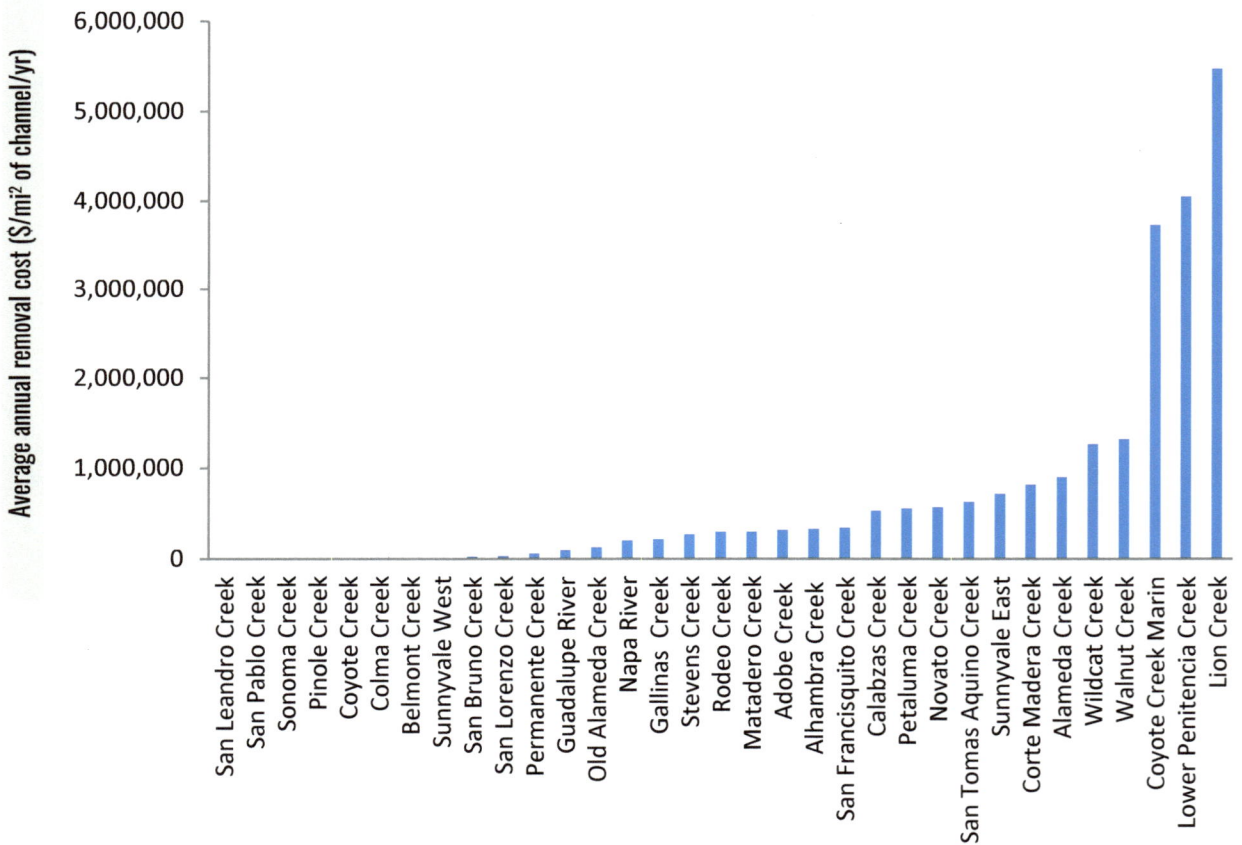

Average annual sediment removal costs (1973-2013). Removal period varies by channel.

(Above) Alameda Creek channel, June 2013, (Courtesy of Dan Rademacher, Creative Commons). *(Below) Eden Landing salt pond restoration, with Alameda Creek channel in center of image, 2014.* (Courtesy of Doc Searles, Creative Commons)

SUMMARY AND SYNTHESIS

Major Findings

For over 50 years, flood control channels have been built and managed in the Bay Area to pass water through the F-T interface to provide protection of people and property in an area where flood danger is caused by diminishing channel gradients near the Bay margin. Data about the sediment supply and depositional processes of these channels has never been compiled before on a regional basis. During this data compilation and analysis effort, we collated data on sediment supply, deposition, removal, grain size and costs for 33 channels. This represents the most comprehensive compilation of sediment data completed to-date for the region. The major findings from this analysis are as follows:

- **Sediment loads in the region are highly variable between years.** In general, the variability of sediment supply to the Bay via flood control channels increases with interannual flow variability and watershed size, and decreases with impervious cover. Highly urbanized watersheds such as Colma Creek tend to have lower sediment load variability due to lower flow variability and more highly managed sediment sources. Larger, less developed watersheds such as Napa River exhibit extreme sediment load variability between years due to much more highly variable runoff between years and more extreme erosional variability. The median interannual variability estimated for the period of 2000-2013 for all 33 watersheds in this study was 3,000-fold.

- **Sediment supply to flood control channels varies greatly across the region in relation to watershed size and other physical characteristics.** Average annual total sediment loads range from <150 tons to >150,000 tons. Average annual sediment yields are also highly variable across the region ranging from 50 and 1,660 tons per square mile per year and reflect geology, climate and land management factors.

- **Sediment deposition data are only sparsely available due to a lack of repeat cross-section and longitudinal profile surveys.** Where data do exist, they help to illustrate a wide variety of depositional characteristics of our channel systems and a variety of management regimes. Some channels are managed to try to maintain the as-built dimensions whereas others have been purposely left to reach an equilibrium sediment deposition condition that local managers have assessed as being of low concern. Most sediment (72%) in recent times (2000-2013) has been removed from tidal reaches and likely has very fine grain size. Flood control channels around the Bay that are trapping coarse sediment (i.e. gravel and larger) in fluvial reaches could be modified to improve coarse sediment transport to the Bay margin.

- **A total of 5.8 million cubic yards (CY) of sediment has been removed periodically from 30 out of 33 flood control channels in the study since 1973**, 30% of which has been removed from 2000 to 2013, a period more representative of the modern policy and management paradigm. More than two-thirds of this sediment is removed from tidal reaches. Over the past several decades, reasonably large and active programs of sediment removal have been occurring in Alameda Creek, Walnut Creek, Petaluma River, Gallinas Creek, Novato Creek, San Tomas Aquino Creek, Napa River, Old Alameda Creek, and Sunnyvale East Channel, a wide spatial distribution around the Bay. Although there is some existing re-use of sediment for restoration, >60% is still being disposed of as waste and not being beneficially re-used for restoration purposes. Most sediment removed between 1973 and 2013 came from channels that were dredged, on average, at least once every 5 years.

- **Sediment removal from flood control channels has cost $111M (not adjusted for inflation) or about $2.8M per year since 1973.** Costs per channel area vary between channels, ranging from $1,225 to $5,459,000 per square mile of channel dredged per year (average cost = $705,000). Currently those costs are being expended by the flood control districts and not shared with the restoration community.

Management Implications

The sediment delivery, removal, and texture data shown here are informative for understanding potential opportunities for habitat restoration. Key management considerations derived from this study are as follows:

- **The extreme spatial and temporal variability in sediment supply from Bay Area creeks provides a very great challenge for management and use of fluvial sediment for wetland restoration,** and suggests that adaptive management will be key for the long-term maintenance of channel conveyance and marsh accretion. While a certain volume of sediment supply cannot be guaranteed for any given year, past records show the range in variability expected over a number of years to decades. However, future climate change may affect the timing and/or intensity of storms, thereby altering watershed sediment supply.

- **In the future, the small, highly urbanized watersheds and those with significant reservoirs will likely continue to deliver relatively low sediment loads,** that will likely be less variable between years. Conversely, less urbanized watersheds may experience greater variability in load delivery between years, including potentially larger total loads.

- **Some channels have established management plans where the volume or timing of sediment removal is prescribed.** Such reliable removal events are more conducive to re-use. However, despite the wide variability of management regimes and removal schedules, more than half of the sediment in flood control channels at the Bay interface has been removed on a frequency of more than once every five years.

- **Sediment deposition and removal is occurring in flood control channels all around the Bay margin,** especially in the larger watersheds in the North Bay counties (Petaluma River, Sonoma Creek, Napa River), East Bay counties (Walnut Creek and Alameda Creek), and South Bay Counties (Coyote Creek and Guadalupe River). This generally wide spatial distribution of sediment supply, deposition, and removal from flood control channels around the Bay creates a higher potential for re-use than if sediment were concentrated in just one area. In contrast, sediment removal for navigation purposes within the Bay is concentrated in the Central Bay, requiring that sediment to be transported to either the North Bay or South Bay to be used in large marsh restoration projects (e.g., in the Napa-Sonoma Marshes, or the South Bay Salt Ponds).

- **Sediment stored in the fluvial portions of our flood control channels tends to be coarser** and represents both a depositional problem for managers as well as an opportunity for high value re-use for specific functions in adjacent baylands. Redesigning our flood control channels so that sediment more effectively passes downstream may be more cost-effective and sustainable over the long-term, in comparison to mechanical sediment removal and transport. In addition, by restoring the sediment transport process, baylands will likely be more resilient to continued future changes.

- **Currently, costs for sediment maintenance are incurred by flood control agencies.** The least cost is usually incurred by disposal in non-beneficial re-use methods (e.g. landfill). However, as sea-level continues to rise and sediment becomes more valuable, the potential for cost sharing with the restoration community will likely increase, thus likely increasing the total amount of sediment that is re-used.

- **The data show that large volumes of sediment are being supplied from the Bay Area watersheds and transported through and deposited within the flood control channels.** Nearly all of this sediment is available or is potentially available to be utilized in tidal marsh restoration or maintenance projects. Future projects should explore linkages between flood control channels and baylands so that both the suspended load and the bedload can be better utilized for supporting natural accretion of marsh plains and for building or nourishing specific coarser grained features (e.g. depositional fans or beaches). Future sea-level rise will require more efficient and comprehensive use of this sediment within baylands in order to effectively protect our cities and infrastructure along the Bay margin.

MULTI-BENEFIT MANAGEMENT MEASURES FOR FLOOD CONTROL CHANNELS AT THE BAY INTERFACE

4

Infrastructure at the shore's edge, Rodeo. (SFEI)

INTRODUCTION

This chapter synthesizes data and findings from Chapters 2 and 3 into conceptual management measures that address sediment delivery to Baylands through natural and mechanical means. Here, we focus on two high-level management measures aimed at delivering sediment to baylands and increasing long-term baylands resilience while helping meet near-term and projected future flood control needs:

- Creek reconnection to baylands - re-establishing the connections flood control channels once had to their adjacent baylands as a way of delivering sediment through natural transport processes to increase accretion rates, thereby increasing bayland resiliency and helping maintain channel capacity

- Local beneficial sediment re-use - using sediment dredged from flood control channels to restore and maintain nearby existing and restored bayland habitats through mechanical placement

We identify measures for the 33 major flood control channels discussed in Chapter 3 based on channel and landscape characteristics, and highlight a few channels as illustrative examples. This information is intended to help the management and restoration communities focus attention early within a channel redesign project when assessing options for integrating habitat improvement and bayland resilience into flood risk management. The measures provide a starting point for multi-benefit channel redesign and do not represent the only applicable measures. The feasibility of the suggested measures could be determined in the design process through detailed studies and technical analyses.

Adaptive management cycle indicating the step where the Flood Control 2.0 project outputs are intended to be used (adapted from Healy et al. 2004).

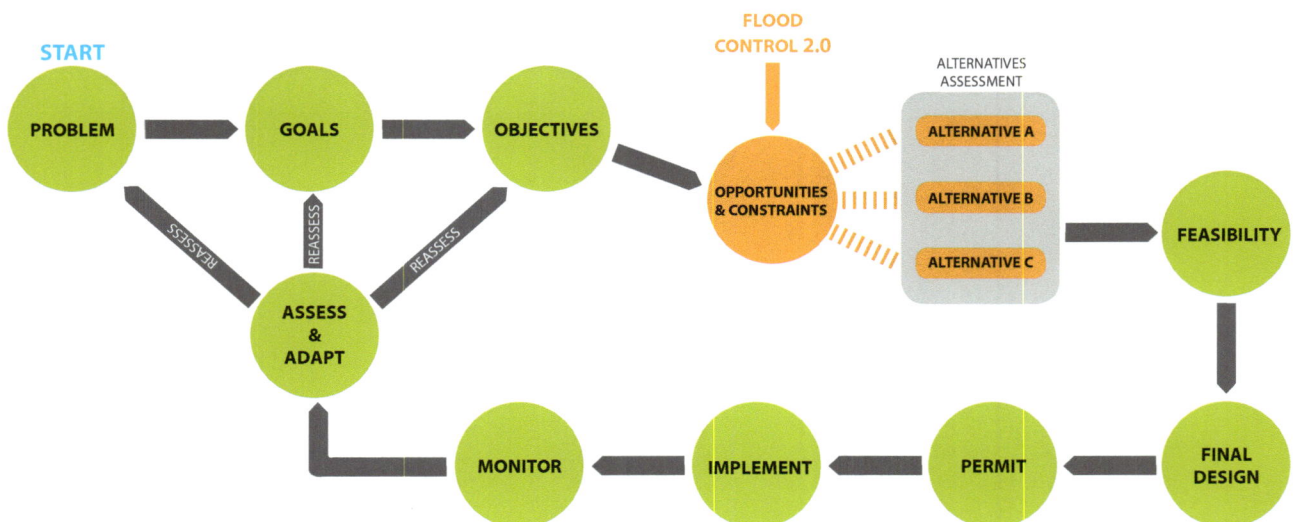

OPPORTUNITIES FOR MULTI-BENEFIT MANAGEMENT

Creek Reconnection to Baylands

Historically, channels draining the watershed surrounding the Bay would deliver freshwater and watershed sediment to baylands, predominantly during flood events. These channels' tidal reaches were also a pathway for fine-grained tidal sediment to get onto baylands. The building of levees (also called dikes) along the tidal reaches has cut-off the adjacent baylands from a regular sediment supply and decreased the tidal prism, which has resulted in in-channel sedimentation issues and decreased flood conveyance capacity that will be exacerbated by sea-level rise. Reconnecting channels to their baylands could relieve many of these issues, thus restoring many of the natural processes and providing the following benefits to ecosystem functions and services:

- Increasing flood storage capacity by setting back levees and spreading fluvial flood waters across the marsh plain, slowing flood velocities and potentially lowering water surface elevations (requiring modeling to verify site specific conditions, and confirming that upstream flood heights do not increase)

- Increasing the long-term delivery of fresh water and sediment to the marsh plain by removing barriers such as levees, maximizing vertical accretion to reduce the potential for marsh drowning with accelerated sea-level rise

- Increasing the tidal prism by setting back or breaching levees so that the slough channel size increases, thereby increasing flood conveyance during low-tide conditions, increasing the channel's ability to transport sediment, and decreasing the overall cost and environmental impact associated with maintenance dredging

- Increasing overall habitat condition for resident wildlife by setting back levees, thereby promoting bayland sediment deposition, establishing salinity gradients, and allowing the exchange of nutrients, food resources, energy and species between the channel and the marsh plain

Local Beneficial Sediment Re-use in Baylands

Much of the sediment currently removed from the flood control channels around the F-T interface is used to cap landfills or is disposed of as a waste product. With the increasing desire to restore lost baylands and with existing baylands being threatened by the predicted inability to maintain their elevation with sea-level rise, the removed sediment could be used to support restored and existing baylands, thereby providing the following benefits to ecosystem functions and services:

- Maintaining and restoring wetland and transition zone habitats with fine sediment, thereby benefitting native fish, birds, and mammals

- Increasing the elevation of subsided or at-risk baylands with fine sediment, thereby providing habitat and helping attenuate waves and reduce coastal flooding risk

- Protecting shorelines with coarse sediment, thereby decreasing erosion risk as large storm frequency increases

- Potentially offsetting dredging costs, as removal and transportation costs could be shared by the party that is re-using the sediment

- Potentially reducing sediment transportation costs, as the removed sediment would be transported a shorter distance to a local re-use site, rather than to a faraway disposal site

(Top right) Levee breach connecting Coyote Creek to former salt pond A19 (March 7, 2006). Photo: Mark Bittner © Pelican Media.

(Bottom right) Pumping of dredged sediment from the Port of Oakland onto the restored Hamilton Field marsh plain (2008). Photo: U.S Army Corp of Engineers.

(Bottom left) A.W. Von Schmidt's dredge in 1884, shown dredging blue clay from mud flats in Oakland Harbor. The pipe extending from the dredger to landfill in distance was used to transport dredge material to create "reclaimed" land.

A. W. VON SCHMIDT'S IMPROVED DREDGING MACHINE.
OAKLAND HARBOR WORKS, CALIFORNIA.
Now working under contract with Colonel George H. Mendell, Corps of Engineers, U.S.A.
MAY, 1884.

METHODS

This analysis focuses on the 33 major flood control channels at the Bay interface described in Chapter 3. To provide channel-specific potential measures focused on sediment delivery that supports habitat restoration and maintenance, we used estimates of a) watershed sediment yield, b) amount of open space adjacent to the channel downstream of the head of tide (i.e., the inland extent of tidal inundation during mean higher high water [MHHW]), and c) historical fluvial-tidal interface type. The approach for determining the estimates used in the analysis is described below.

Sediment yield (tons/mi^2/yr) was estimated for WYs 2000-2013 for the portion of each watershed downstream of any major water supply dams. First, we used streamflow and sediment gauge records where available and regional regressions in watersheds where data was not available to arrive at the estimated climatically averaged total watershed sediment load for the recent past. Next, total sediment load was normalized by watershed area downstream of dams (as appropriate) to arrive at average annual sediment yield. We then divided the watersheds into three sediment yield categories: high (>900 tons/mi^2/yr), medium (300-900 tons/mi^2/yr), and low (<300 tons/mi^2/yr). See Chapter 3 for more detail regarding the methods for determining watershed sediment yield and the yield categories.

Each channel was also classified based upon the amount of "available space," or extent of undeveloped lands, adjacent to the channel that has the potential to be inundated by watershed flood flows and tidal flows if the historical channel-bayland connection is re-established. We classified the channels as having "space" or "no space" qualitatively by examining recent aerial photographs (e.g. NAIP 2012) and land use maps (e.g., ABAG 2006). Channels classified as having "space" have relatively large non-urban low-lying upland, tidal marsh, or diked marsh areas immediately adjacent to the flood control channel either at head of tide or along the tidal reach downstream. This includes channels with vast amounts of undeveloped area (e.g., Novato Creek, Sonoma Creek, Napa River, Walnut Creek, Alameda Creek, Coyote Creek, and Guadalupe River) as well as those with smaller pockets of undeveloped areas (e.g., Wildcat Creek, Coyote Creek Marin, San Lorenzo Creek, San Francisquito Creek, Pinole Creek, and Alhambra Creek).

For the channels that were classified as having "no space," we wanted to further explore the physical characteristics that were controlling sediment deposition in the channel. The F-T interface type findings from Chapter 2 were utilized to illustrate how sediment was historically transported through the channels and delivered to baylands. A handful of channels were historically disconnected, ending in distributaries inland where their water and sediment would spread out across the alluvial plain. Sometimes this occurred relatively close to the baylands margin (< 1 mile distance, e.g., Coyote Creek in Marin County), while at other times it was much closer to the hills (e.g., Stevens Creek in Santa Clara County). Comparing the location of the historical distributary to the current location of dominant sediment deposition was informative for channels of this type, and highlighted watersheds where the physical characteristics (e.g., channel gradient, stream power) could be driving a portion of the sediment deposition process.

Example channel with adjacent undeveloped land and the potential for creek-baylands reconnection (Sonoma Creek, top), and example channel too constrained by development for channel-baylands reconnection (Colma Creek, bottom).

Summary of regional channel physical information and baylands enhancement measures for the 33 channels assessed.

Creek name	County	Classification of adjacent space	Watershed sediment yield (2000-2013) free-flowing area	Historical F-T interface type	Distance of historical channel terminus from Bay margin	Priority baylands enhancement measure
Sunnyvale East Channel	Santa Clara	Space	Low	New channel	-	Creek Connection
Sunnyvale West Channel	Santa Clara	Space	Low	New channel	-	Creek Connection
Alameda Creek	Alameda	Space	Low	Tidal marsh channel	-	Creek Connection
Novato Creek	Marin	Space	Low	Tidal marsh channel	-	Creek Connection
Coyote Creek	Santa Clara	Space	Low	Tidal marsh channel w/ natural levee	-	Creek Connection
Guadalupe River	Santa Clara	Space	Low	Tidal marsh channel w/ natural levee	-	Creek Connection
Alhambra Creek	Contra Costa	Space	Medium	Bay	-	Creek Connection
Pinole Creek	Contra Costa	Space	Medium	Tidal marsh channel	-	Creek Connection
San Pablo Creek	Contra Costa	Space	Medium	Tidal marsh channel	-	Creek Connection
Napa River	Napa	Space	Medium	Tidal marsh channel w/ natural levee	-	Creek Connection
Old Alameda Creek	Alameda	Space	Medium	Tidal marsh channel w/ natural levee	-	Creek Connection
San Francisquito Creek	San Mateo	Space	Medium	Tidal marsh channel w/ natural levee	-	Creek Connection
San Tomas Aquino	Santa Clara	Space	Medium	Tidal marshland	-	Creek Connection
San Lorenzo Creek	Alameda	Space	Medium	Tidal marshland w/ natural levee	-	Creek Connection
Adobe Creek	Santa Clara	Space	Medium	Disconnected	Far	Creek Connection
Calabazas Creek	Santa Clara	Space	Medium	Disconnected	Far	Creek Connection
Matadero Creek	Santa Clara	Space	Medium	Disconnected	Far	Creek Connection
Permanente Creek	Santa Clara	Space	Medium	Disconnected	Far	Creek Connection
Petaluma River	Sonoma	Space	High	Tidal marsh channel	-	Creek Connection
Wildcat Creek	Contra Costa	Space	High	Tidal marsh channel	-	Creek Connection
Sonoma Creek	Sonoma	Space	High	Tidal marsh channel w/ natural levee	-	Creek Connection
Gallinas Creek	Marin	Space	High	Tidal marshland	-	Creek Connection
Walnut Creek	Contra Costa	Space	High	Tidal marshland	-	Creek Connection
Coyote Creek Marin	Marin	Space	High	Disconnected	Near	Creek Connection
Stevens Creek	Santa Clara	Space	High	Disconnected w/ natural levee	Far	Creek Connection
San Leandro Creek	Alameda	No Space	Low	Tidal marsh channel	-	Sediment Re-use
Corte Madera Creek	Marin	No Space	Medium	Tidal marsh channel	-	Sediment Re-use
Lower Penitencia Creek	Santa Clara	No Space	Medium	Tidal marsh channel	-	Sediment Re-use
San Bruno Creek	San Mateo	No Space	Medium	Disconnected	Near	Sediment Re-use
Colma Creek	San Mateo	No Space	High	Tidal marsh channel	-	Sediment Re-use
Rodeo Creek	Contra Costa	No Space	High	Tidal marsh channel	-	Sediment Re-use
Belmont Creek	San Mateo	No Space	High	Tidal marshland	-	Sediment Re-use
Lion Creek	Alameda	No Space	High	Tidal marshland	-	Sediment Re-use

RESULTS

Creek Reconnection to Baylands

Using the extent of undeveloped lands adjacent to the channels, we identified creeks with the potential for channel reconnection for bayland habitat support. Of the 33 flood control channels that were considered in the study, 25 channels were classified as having "space," with at least some undeveloped land adjacent to the channel. Reconnecting these creeks to existing or restored adjacent baylands would allow sediment from the watershed and sediment scoured from the channel to be distributed across the baylands area that has been "opened" to these flows.

However, not all creeks are the same, and the magnitude of impact of this measure may vary, especially when considering the total amount of space available and the amount of sediment within the system. Creeks that have abundant adjacent space (e.g., baylands, diked baylands, or undeveloped upland) will likely have a wider variety of options for reconnection, and could support a much greater suite of benefits due to the reconnection. Alternatively, creeks that only have a small "pocket" of space available might also benefit from this action, but the magnitude of impact might be much smaller. In addition, the amount of sediment provided from the watershed will also have an effect; sites with higher amounts of sediment may have faster vertical accretion rates, may support larger bayland areas, or may be appropriate for a wider suite of uses. See Collins (2006) for an assessment of potential marsh accretion rates based on adjacent watershed sediment supply for marshes throughout the region.

Here, we highlight three examples of creeks with the potential for channel reconnection to baylands to illustrate the potential benefits and considerations for coordination with other restoration actions. We focus on creeks that represent the low and high ends of the average annual sediment yield range, and that represent areas with extensive and more limited "available space." For each creek shown, we provide maps of contemporary habitat and the amount of "elevation capital" (i.e., the elevation of land in relation to the tidal frame) for a 5 ft (1.5 m) rise in mean sea-level, which is a reasonable estimate of the mean sea-level rise by the end of the century (NRC 2012). Assessing elevation capital helps in understanding the overall need for sediment to maintain existing and restored baylands over the long-term.

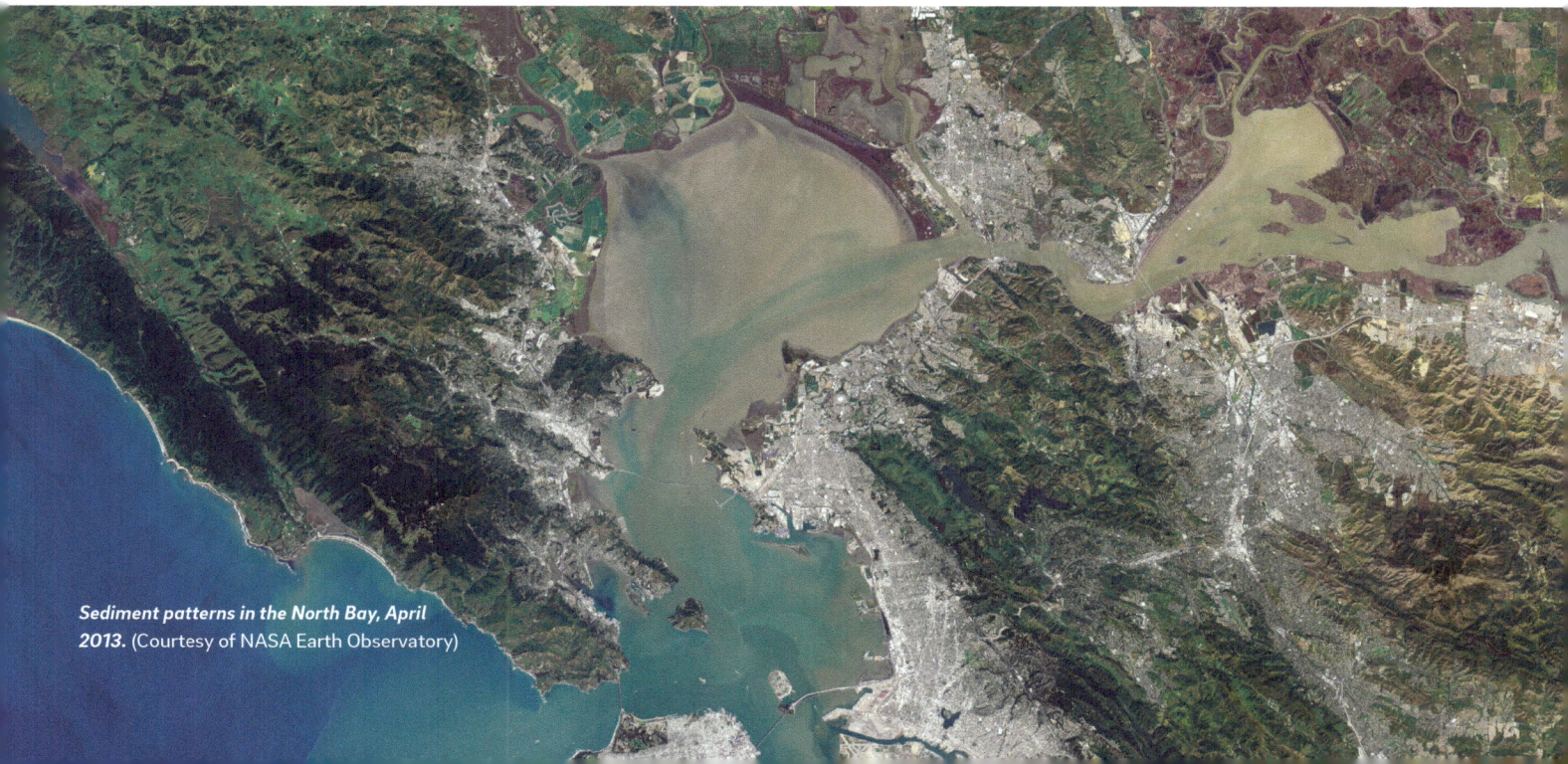

Sediment patterns in the North Bay, April 2013. (Courtesy of NASA Earth Observatory)

Napa River

Petaluma River

Sonoma Creek

A

Novato Creek

Gallinas Creek

Alhambra Creek

Pinole Creek

C

Wildcat Creek

San Pablo Creek

B

Walnut Creek

Coyote Creek

San Lorenzo Creek

Old Alameda Creek

Estimated average annual sediment yield (tons/mi²/yr) for channels with considerable undeveloped adjacent land and the potential for creek-baylands reconnection. The letters correspond to the example channels shown in more detail on page 51, and discussed on page 52.

Alameda Creek

SEDIMENT YIELD (2000-2013)

△ High (>900 tons/mi²/yr)

△ Medium (300-900 tons/mi²/yr)

△ Low (<300 tons/mi²/yr)

San Francisquito Creek

Matadero Creek

Adobe Creek

Permanente Creek

Stevens Creek

Sunnyvale W. Channel

Coyote Creek

Calabazas Creek

Guadalupe Creek

San Tomas Aquino Creek

Sunnyvale E. Channel

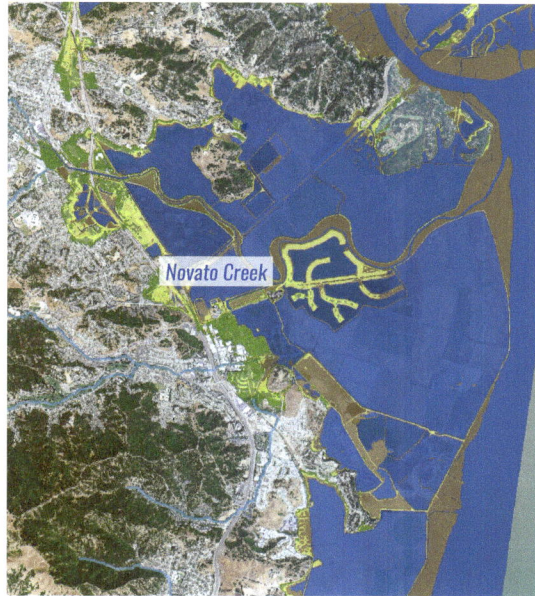

A

1:120,000

Novato Creek

Novato Creek

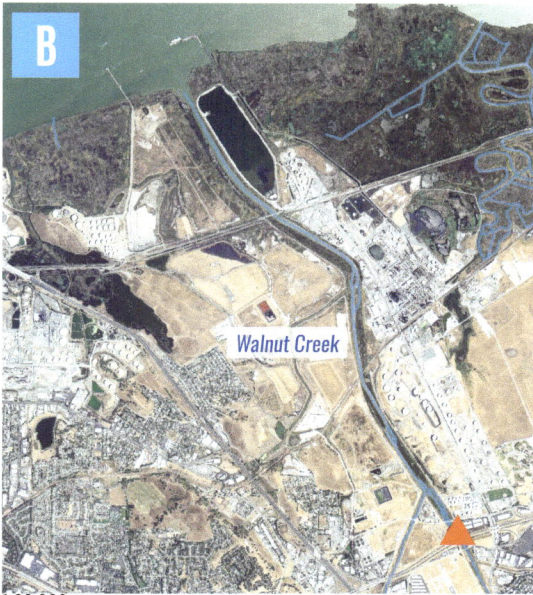

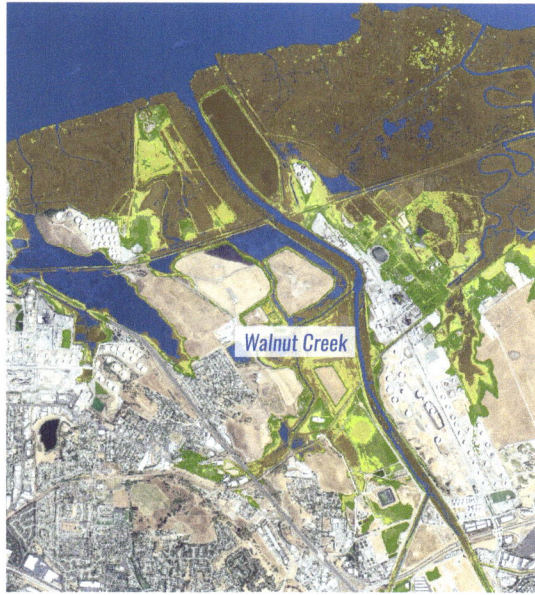

B

1:80,000

Walnut Creek

Walnut Creek

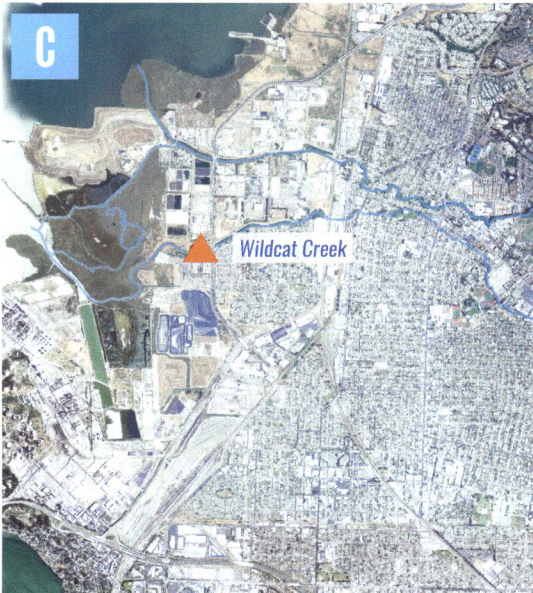

C

1:80,000

Wildcat Creek

Wildcat Creek

SEDIMENT YIELD

▲ High

△ Low

POTENTIAL FUTURE HABITAT TYPE

- Subtidal
- Intertidal Mudflat
- Low Marsh
- Mid Marsh
- High Marsh
- Transition

51

A ***Novato Creek*** is a location with abundant adjacent undeveloped space but a relatively low average annual watershed sediment yield. Creek reconnection may provide some improvement in flooding risks and would allow sediment delivery to the marsh that would be recreated within the historical baylands footprint. However, as the adjacent land is currently at a very low elevation (as indicated by the elevation capital map), natural delivery of watershed and tidal sediment may need to be augmented to allow those restored bayland areas to keep pace with sea-level rise. These concepts are further developed in the Novato Creek Baylands Vision (SFEI-ASC 2015) that was developed as part of the Flood Control 2.0 project.

B ***Walnut Creek*** is a location also with abundant adjacent undeveloped space and a relatively high average annual watershed sediment yield. Reconnecting the creek with the adjacent baylands could not only potentially provide some reductions in flooding risks, but also an opportunity for significant bayland habitat creation and maintenance over the long-term. The relatively high sediment supply could be effective at nourishing those areas shown on the elevation capital map that are anticipated to convert to mudflat without more sediment input. These concepts are further developed in the Walnut Creek Baylands Vision (SFEI-ASC 2016) that was developed as part of the Flood Control 2.0 project.

C ***Wildcat Creek*** is a location with a relatively low amount of adjacent available undeveloped space but a relatively high average annual watershed sediment yield. Creek reconnection could help reduce flooding risks and allow for the creation and maintenance of small, confined marsh areas within the historical bayland footprint. Watershed sediment would need to be directed toward these areas though strategic connection points and delivery would need to be concentrated on the lowest elevation areas (e.g., the upland parcels shown at mudflat elevation on the elevation capital map).

Local Beneficial Sediment Re-use in Baylands

Using the extent of undeveloped lands adjacent to the channels and watershed sediment supply, we identified creeks where beneficial sediment re-use appears to be the most viable option for bayland habitat support. Of the 33 flood control channels that were considered in the study, we identified eight creeks with "no space," or no adjacent undeveloped land that could be utilized for creek reconnection. In the past, the majority of sediment removed from these channels for maintaining conveyance capacity was taken to landfills or disposed of in aquatic environments, with very limited local re-use. Mechanical sediment removal from these channels will likely need to continue in the future and may need to increase in frequency as sea-level rises and the channels' tidal reaches become more depositional. The sediment removed from these channels could be re-used locally, targeting projects that need sediment to increase the resiliency of existing or restored baylands. It's likely that channels with a wide range of particle sizes could have the most options for beneficial re-use because they could help meet the needs of efforts that need finer sediment for building up tidal marsh surface elevations and projects that need coarser sediment for building beaches to protect baylands from shoreline erosion. The SediMatch online tool that was developed as part of the Flood Control 2.0 project can be used to make the necessary connections between the dredging community and the bayland restoration and management community.

Here, we highlight two examples of creeks with the potential for beneficial sediment re-use. We focus on creeks that represent a range of average annual sediment yields and physical settings that dictate the options for reusing dredged sediment.

(Below left) Wildcat Creek after heavy rains, March, 2011. (Courtesy of Nick Fullerton, Creative Commons)

(Below right) Novato Creek after heavy rains, December, 2005. (Courtesy of Jessica Merz, Creative Commons)

Estimated average annual sediment yield (tons/mi²/yr) for channels with little to no undeveloped adjacent land where the focus should be on beneficial re-use of sediment on adjacent tidal habitats. The letters correspond to the example channels shown in more detail on page 55, and discussed on page 56.

A

B

Rodeo Creek

Corte Madera Creek

Lion Creek

San Leandro Creek

Colma Creek

San Bruno Creek

Belmont Creek

Lower Penitencia Creek

SEDIMENT YIELD (2000-2013)

▲ High (>900 tons/mi²/yr)

▲ Medium (300-900 tons/mi²/yr)

▲ Low (<300 tons/mi²/yr)

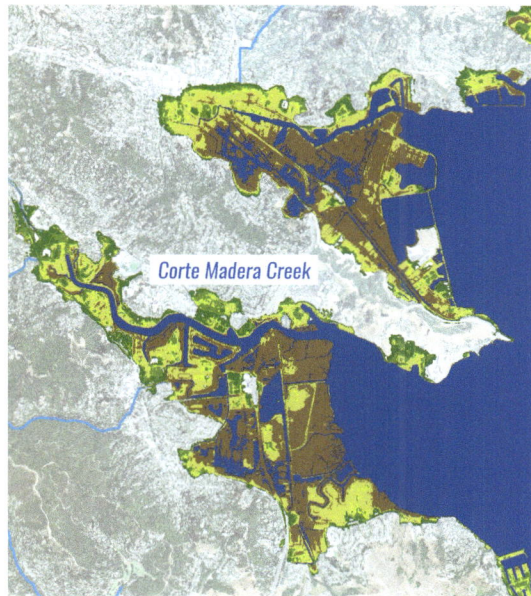

1:100,000

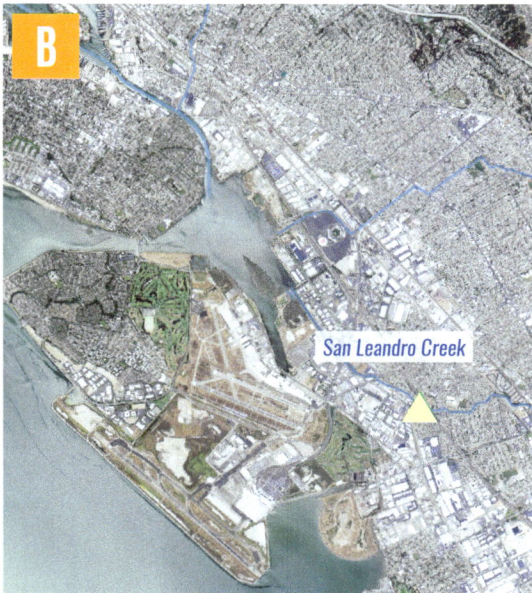

1:120,000

SEDIMENT YIELD

▲ Medium

△ Low

POTENTIAL FUTURE HABITAT TYPE

■ Subtidal
■ Intertidal Mudflat
■ Low Marsh
■ Mid Marsh
■ High Marsh
■ Transition

Example channels where the focus should be on beneficial sediment re-use. The maps on the right indicate each site's "elevation capital" as habitat types that would be present if all areas were open to the tide and mean sea level increased by 5 ft without an associated increase in sediment supply.

A **Corte Madera Creek** is a fairly unique flood control channel, in that it has development right up to the channel banks for its entire length downstream of head of tide, but it also has a significant area of marsh adjacent to the mouth that is providing important flood protection for the shoreline. Sediment manipulation is common; the flood control channel accumulates sediment and sediment removal is conducted regularly to ensure flood capacity. In addition, the nearby Larkspur Ferry channel is also regularly dredged to support navigation. A recent study by the San Francisco Bay Development and Conservation District (BCDC) suggests that the baylands, including the mudflats and tidal marshes, will not survive as sea-level rises without additional sediment inputs (BCDC 2013). The study recommends stabilizing the marsh edge with coarse sediment, building up the mudflat with fine sediment, and creating a gently sloping tidal-terrestrial transition zone slope that allows the marsh to migrate over time. With additional detailed study and support from the regulating agencies, the removed sediment could be re-used to address these recommendations and also support vertical marsh accretion.

Presently, there is only limited information on grain size available in the dredge material and there is historic grain size information on the fluvial load at the gauge site at Ross dating from WY 1978-1980 but no recent data. Since these end uses are size related, the future collection of grain size data throughout the system will be important.

B **San Leandro Creek** currently accumulates primarily tidal sediment. The upper portion of the watershed is impounded by a dam, thus the amount of fluvial sediment supplied to New Marsh and Arrowhead Marsh at the mouth of the creek is relatively low compared to historical loads. Due to marsh wildlife species concerns, namely habitat for Ridgway's rails, local re-use around the mouth of San Leandro Creek may not be feasible. However, sediment removed from San Leandro Creek could be used to nourish existing and restored baylands and transition zone features in nearby areas. For example, the East Bay Dischargers Authority (EBDA) is currently examining alternatives for discharging treated wastewater through a seepage slope on a horizontal levee behind Roberts Landing and Ora Loma marshes, thereby creating habitat for resident wildlife and marsh migration space as sea-level continues to rise (Beyeler et al. 2015). The sediment removed from San Leandro Creek could be used to build that horizontal levee and maintain it over time.

Ridgway's rail, January 2016, near Pt. Isabel. (Courtesy of Becky Masubara, Creative Commons)

Hybrid Combining Both Measures

The physical setting of several South Bay creeks suggests prioritizing a hybrid approach that combines creek reconnection and beneficial sediment re-use for bayland habitat support. These creeks flow through undeveloped diked baylands, suggesting there could be opportunities for connecting the creek to restored baylands. However, the channels were also historically disconnected from the tidal environment (as discussed in Chapter 2), with their historical channel terminus miles away from the Bay. Despite channel modifications aimed at rapidly moving flood waters through these channels and out to the Bay, the underlying landscape slope on the South Bay alluvial plain has remained relatively static and the channels still have a tendency to lose stream power and deposit sediment around their historical channel terminuses due to the change in gradient from the watershed to the adjacent baylands. Thus, since sediment will likely continue to accumulate in these channel locations regardless of any efforts to connect the creeks to baylands downstream, beneficial re-use of the dredged sediment should also be considered.

Below we highlight one example where there's the potential for a channel management approach that combines creek reconnection and sediment re-use. The example chosen is considered to be representative of the conditions that exist at the other South Bay channels that also have the potential for a combination of both measures.

EXAMPLE: ADOBE CREEK IN SANTA CLARA COUNTY

Over the past several decades, the Santa Clara Valley Water District has removed approximately 59,000 CY of sediment from the Adobe Creek flood control channel (Sara Duckler, SCVWD, personal communication). Of this, approximately 70% came from the tidal reach downstream of head of tide and approximately 30% came from the fluvial reach between head of tide and the highly urbanized region upstream where the channel historically terminated during low flow conditions. Connecting the tidal reach to restored baylands around the channel mouth would enable regular fluvial and tidal inundation of the baylands, and would promote channel scouring and decreased sediment deposition in the tidal reach due to increased tidal prism. However, it's likely that the increased channel capacity due to tidal scour would not propagate that far into the fluvial reach, and channel dredging around the historical channel terminus would still be necessary to maintain flood conveyance capacity. The dredged fluvial sediment likely ranges in size from silt to cobble and could therefore be used for a variety of bayland restoration and maintenance applications, including marsh plain accretion with the finer sediment and marsh edge protection with the coarser sediment.

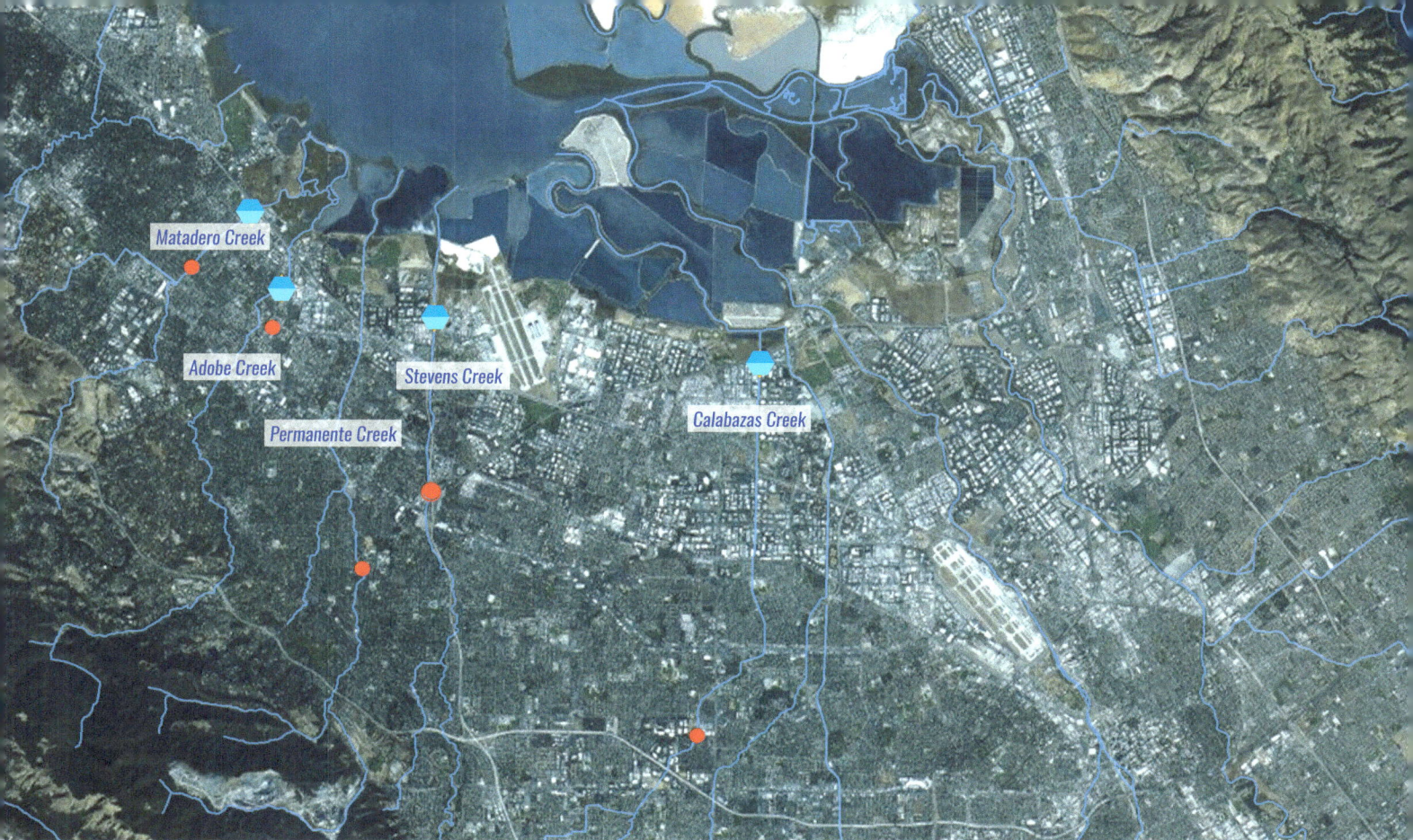

Five South Bay channels with the physical setting appropriate for a hybrid approach that combines creek-bayland reconnection and beneficial sediment re-use (top), and sediment removal locations over the past three decades for one example channel (Adobe Creek, bottom).

Head of tide location

Historical channel terminus

Sediment removal reach (1980-2012)

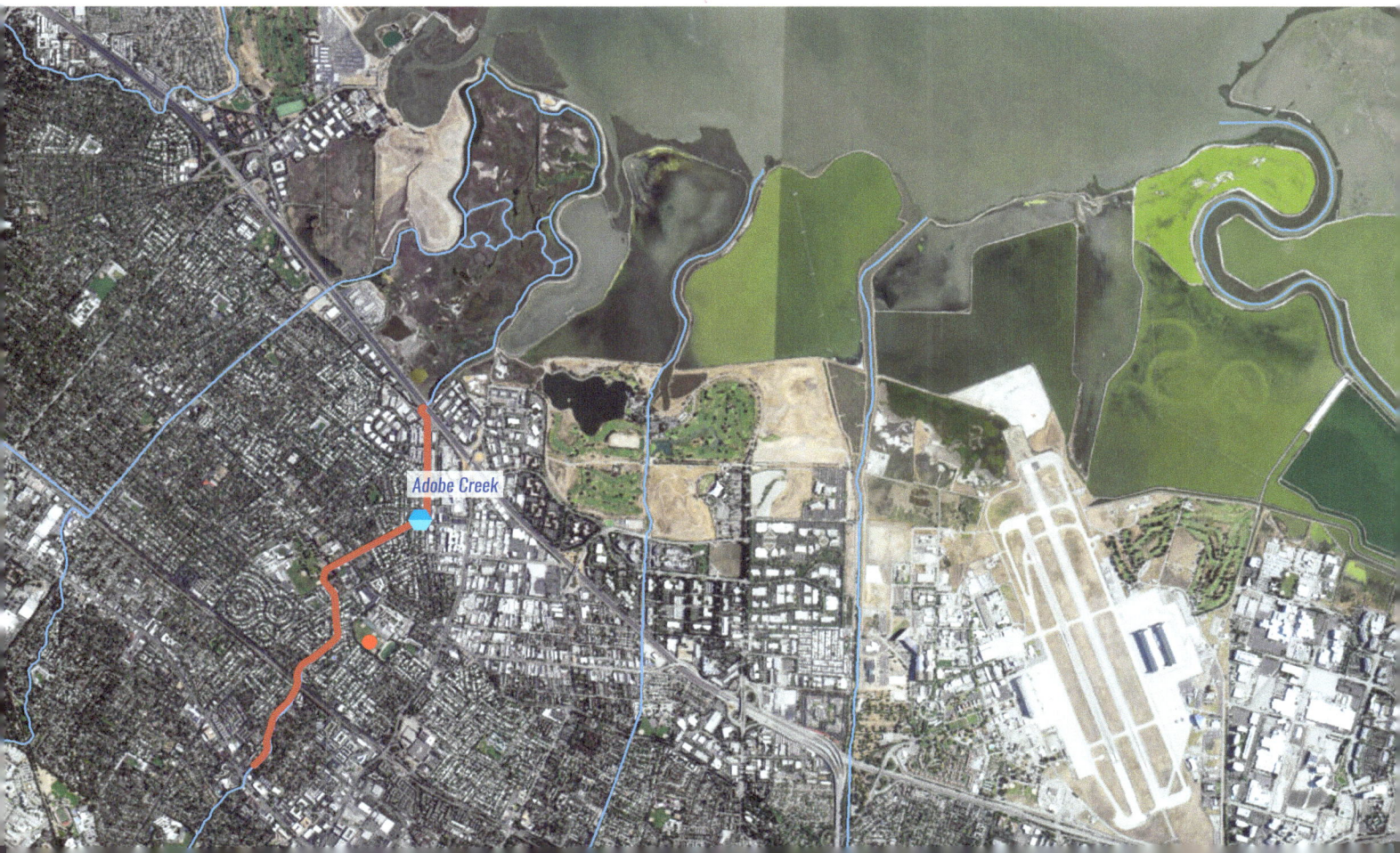

Mouth of Adobe Creek, 2013.
(Courtesy of Edward Roofs, Creative Commons)

SUMMARY AND SYNTHESIS

Major Findings

Faced with challenges of excess sediment deposition in flood control channels near the Bay and adjacent baylands that need sediment to survive under a rising sea level, the San Francisco Bay land management and restoration communities are now perceiving the sediment that passes through and gets trapped in flood control channels as a valuable habitat restoration resource. In this chapter, we combined the channel morphology and sediment information presented in the previous chapters, along with additional information on landscape characteristics, to illustrate multi-benefit channel management measures that could provide sediment to bayland habitats (through both natural and mechanical means) within the context of improving flood risk management. We considered two management measures; creek reconnection to baylands, which entails re-establishing the connections flood control channels once had to their adjacent baylands, and local beneficial sediment re-use, which entails using sediment dredged from flood control channels to build and maintain nearby bayland habitats. The major findings from the analysis are as follows:

- **For 25 of the 33 channels considered, channel reconnection was identified as a priority measure for getting sediment to bayland habitats due to the presence of undeveloped land adjacent to the channels' tidal reach that could be reconnected**

to river and tidal flows. Creeks with abundant adjacent space would likely have the widest variety of options for reconnection and could support the greatest suite of ecosystem services. Creeks with a high watershed sediment supply could likely support larger bayland areas that have accretion rates needed for long-term resilience.

- **For the other eight channels, beneficial sediment re-use was identified as a priority measure** for getting sediment to bayland habitats because the channels' tidal reaches are so highly constrained by in-channel infrastructure and land development that mechanical sediment removal will likely continue to be the most viable sediment management option. Instead of being used as landfill cover or disposed of as a waste product, this sediment could be re-used locally for projects that need sediment to increase the resiliency of existing or restored baylands. Channels with a wide range of particle sizes may conceivably have the most options for beneficial re-use.

- **Of the 25 channels having channel reconnection as the priority measure, five channels located in the South Bay have a landscape setting that suggest an approach for moving sediment to bayland habitats that also includes beneficial sediment re-use.** Historically, these channels were disconnected from the tidal environment, with their terminus on the alluvial plain miles inland from the Bay. These creeks now have a permanent connection to a tidal channel with undeveloped land adjacent to the channel's tidal reach, suggesting there could be an opportunity to implement the creek reconnection measure. However, these channels also currently trap watershed sediment upstream of the tidal zone in the area of the historical channel terminus. Sediment removal would likely need to continue in this area even if channel reconnection occurred downstream, suggesting sediment re-use could also be a priority measure.

Application

The management measures given here for flood control channels around San Francisco Bay are intended to provide ideas about potential multi-benefit management opportunities for bringing habitat restoration into flood risk management that could be explored at the start of a channel re-design project. These measures were assigned to the channels based on a high level understanding of historical and contemporary sediment yield and in-channel sediment deposition dynamics, and a qualitative assessment of the amount of potential land that could be reconnected to the channels along their tidal reaches. However, the feasibility of these measures for each channel is not known and would need to be determined through detailed constraints assessments and technical analyses. After determining that the measures would not negatively impact flood protection, key site-specific constraints that would need to be considered include site conditions (e.g., soil quality), necessary infrastructure modifications, access to property that would be reconnected to the channel, permitting challenges, and overall cost.

King tide at Sonoma Creek, January 9, 2013. (SFEI)

RECOMMENDATIONS

5

SUMMARY OF FINDINGS

The analyses presented in this report explored morphologic change and sediment dynamics in flood control channels at fluvial-tidal (F-T) interface around San Francisco Bay, and provide multi-benefit management concepts aimed at bringing habitat restoration into flood risk management. The major findings from the analyses are as follows:

- Historically, the dominant F-T interface types around San Francisco Bay were creeks that connected directly to the Bay, creeks that connected to a tidal channel network, creeks that drained onto tidal marshland, and creeks that were unconnected to the tidal environment (except during large floods). The major drivers controlling the historical interface type include stream power during floods (i.e., product of channel slope and flow discharge) and watershed sediment supply. Over the past 200 years, most channels have been altered for land reclamation and flood control so that they have a permanent connection to a tidal channel that flows through diked baylands or bay fill, or have been routed underground or filled in completely at the historical F-T interface location.

- Sediment supply to and removal from flood control channels that drain to San Francisco Bay vary considerably. Current estimated average annual sediment yields for 33 major flood control channels range from 50 to 1,660 tons/mi^2/yr and reflect differences in watershed geology, climate and land management factors. Highly urbanized watersheds tend to have lower sediment yields and lower interannual sediment yield variability due to lower flow variability and more highly managed sediment sources. Over the past four decades, approximately two-thirds of the 5.8 million cubic yards of sediment removed from 30 of the major Bay Area flood control channels came from tidal reaches (i.e., reaches downstream of head of tide). Most of the sediment came from channels that were dredged, on average, at least once every five years, and most was taken to landfills or disposed of as a waste product. In addition, sediment removal from flood control channels since 1973 has cost $115M (not adjusted for inflation), with individual channel costs ranging from $1,225 to $5,459,000/mi^2 of channel dredged/yr.

- The findings from the F-T interface assessment and sediment analysis were used to highlight conceptual management measures that address sediment delivery to baylands through natural and mechanical means. Of the 33 flood control channels considered, 25 were identified as having the potential for channel reconnection as a means of getting sediment to bayland habitats because they have undeveloped land adjacent to the channels' tidal reach that is currently protected by levees but could be reconnected to river and tidal flows. For the other eight channels, beneficial sediment re-use appears to be the most viable option for getting sediment to bayland habitats because they are highly constrained by in-channel infrastructure and land development along the channels' tidal reach. The physical setting of five South Bay creeks considered suggests a hybrid approach that includes both creek reconnection and beneficial sediment re-use could be an effective management approach for getting sediment to bayland habitats.

KEY RECOMMENDATIONS

Although flood control agencies and other entities currently collect data that are useful for developing multi-benefit management strategies that benefit people and wildlife, additional data are necessary, particularly data related to sediment. Specifically, more information is needed on the quantity and quality of sediment that deposits in and travels through the F-T transition and downstream tidal reaches of flood control channels to help us better understand the sediment that is available for tidal habitat restoration projects, now and into the future. Below, we provide recommendations for data collection efforts and quantitative analysis focused on sediment that would help with the development of multi-benefit management strategies. These recommendations are intended to augment recommendations coming from other efforts in the region focused on sediment science as it relates to long-term bayland management under changing climatic conditions (e.g., BCDC Sediment Science Strategy).

Watershed sediment supply

There are currently very few measurements of watershed suspended load and bedload being taken at the head of flood control channels that drain to the Bay. These data are necessary for not only improving our estimates of overall average annual and episodic watershed sediment yields, but also improving our understanding of the portion of that sediment depositing in flood control channels. Better information on watershed sediment delivery would help identify channels where excess watershed sediment load drives deposition and loss of flood conveyance, and would help inform redesign options for improving sediment transport capacity.

RECOMMENDATION

Continuous suspended load and episodic bedload data (both mass/time and size distribution) should be collected for at least the 33 major flood control channels discussed in Chapters 3 and 4. Data collection should follow the standardized techniques developed by the USGS and the data should be made publicly available through a regional data center and/or through the USGS NWIS data portal.

In-channel sediment storage

As with watershed sediment supply data, sediment storage data for flood control channels that drain to the Bay is currently lacking. In particular, very few flood control channels have ongoing regular channel cross-section or longitudinal profile surveys to document the change in channel elevation, and the associated change in stored sediment volume, over the long-term and directly following major storms or dredging events. These data are needed to determine long-term sediment storage trends for individual channels (i.e., whether channels are actively aggrading, incising, or in equilibrium) and can be used with watershed sediment supply data to determine the portion of watershed sediment that gets trapped in flood control channels. This type of information is essential for developing appropriate redesign options aimed at improving sediment transport capacity and routing sediment to adjacent tidal habitats.

RECOMMENDATION

Channel cross-section and longitudinal profile surveys of flood control channels should be conducted regularly for at least the 33 major flood control channels discussed in Chapters 3 and 4. The survey frequency should be at least once every three to five years, with surveys also occurring directly after major storms (e.g., 2-year floods and larger). In addition, grain size distribution data for channel bed sediment should be collected during the topographic surveys. Standardized surveying and grain size data collection techniques should be followed and the data should be made publicly available through a regional data center.

Sediment removal

Currently, there is often little specific information about the location, volume, texture, and ultimate fate of sediment dredged from flood control channels for individual dredging events, as well as the total dredging cost including coordination and permitting. Data pertaining to the location, volume, grain size, and fate of dredged sediment are needed for many channels to help elucidate the drivers for excess sediment accumulation problems and the amount of sediment that could be available for a range of tidal habitat restoration projects (e.g., fine sediment appropriate for building marsh plains, coarse sediment appropriate for building beaches). Data pertaining to the total cost of dredging is useful for understanding the actual cost of channel dredging (i.e., the cost including all of the work preparing for a dredging event and getting all necessary permits approved) and identifying long-term channel management costs that could be avoided by implementing a channel design that improves sediment transport capacity and decreases the need for dredging.

RECOMMENDATION

Information from all sediment removal events should be entered into a regional database. Data collected should include the removal location, sediment volume, sediment grain size, sediment fate, and costs for individual dredging events. These data should be collected using standardized methods and the data should be made publicly available through a regional data center.

Future conditions

Although some creeks have information about projected impacts of land use changes and climate change on sediment delivery and deposition rates, more detailed information is needed throughout the region. Detailed numerical modeling of watershed sediment yield and sediment deposition in the channel reaches around the fluvial-tidal transition and downstream should be conducted using the best available information on planned near-term land use and channel management, precipitation dynamics (e.g., large storm frequency), and Bay water surface elevation (e.g., mean tide and storm surge). Understanding how the combination of land use, channel management, and climate change could affect sediment dynamics could help identify situations where watershed and/or channel management approaches will need to be modified to avoid excess sediment deposition, and will allow us to get a better handle on the amount of watershed sediment that could be delivered to baylands and the Bay in the future.

> **RECOMMENDATION**
>
> There should be more numerical modeling of climate change scenarios for sediment transport to and deposition within at least the major 33 flood control channels discussed in Chapters 3 and 4. The modeling should be done using established methods and the results should be made publicly available through a regional data center.

Additional sediment sources

There are currently dozens, if not hundreds, of small channels that drain small catchments and flow through stormwater pipes before discharging into the Bay. Individually, these pipes contribute very little to the total sediment load to the Bay. However, we do not know if these pipes contribute a significant portion of the total sediment load collectively. Quantifying the sediment load from these pipes could help improve estimates of total sediment load to the Bay, which will ultimately help with developing strategies aimed at managing the Bay sediment supply to promote long-term bayland resilience.

> **RECOMMENDATION**
>
> The conversation about sediment delivery to the Bay should be extended beyond flood control agencies to include city agencies who manage the stormwater infrastructure. An estimate of sediment contributions supplied to the Bay from the many hundreds of stormwater pipe outfalls should be supported by a regional monitoring program. Sediment data should be collected using standardized methods and the data should be made publicly available through a regional data center.

Table 1. Estimated average annual sediment load for the 33 channels assessed.

Name	County	Average Annual Sediment Load 1957-2013 (tons)	Average Annual Sediment Load 2000 - 2013 (tons)
Adobe Creek	Santa Clara	11,297	7,682
Alameda Creek	Alameda	127,727	93,507
Alhambra Creek	Contra Costa	16,814	13,397
Belmont Creek	San Mateo	3,983	3,170
Calabazas Creek	Santa Clara	14,325	9,740
Colma Creek	San Mateo	31,217	18,845
Corte Madera Creek	Marin	10,619	6,478
Coyote Creek	Santa Clara	10,342	7,032
Coyote Creek Marin	Marin	3,039	1,854
Guadalupe River	Santa Clara	12,834	8,727
Gallinas Creek	Marin	1,912	1,166
Lion Creek	Alameda	4,187	3,419
Lower Penitencia Creek	Santa Clara	21,684	14,744
Matadero Creek	Santa Clara	11,280	7,670
Napa River	Napa	176,366	112,824
Novato Creek	Marin	4,207	3,586
Old Alameda Creek	Alameda	13,690	11,180
Permanente Creek	Santa Clara	22,337	12,400
Petaluma River	Sonoma	36,225	49,145
Pinole Creek	Contra Costa	5,616	5,513
Rodeo Creek	Contra Costa	11,936	9,510
San Bruno Creek	San Mateo	4,603	3,663
San Francisquito Creek	San Mateo	23,119	18,398
San Leandro Creek	Alameda	663	559
San Lorenzo Creek	Alameda	16,851	13,761
San Pablo Creek	Contra Costa	8,770	6,988
San Tomas Aquino Creek	Santa Clara	28,560	19,420
Sonoma Creek	Sonoma	112,487	152,606
Stevens Creek	Santa Clara	25,021	17,014
Sunnyvale East Channel	Santa Clara	490	333
Sunnyvale West Channel	Santa Clara	207	141
Walnut Creek	Contra Costa	181,039	144,255
Wildcat Creek	Contra Costa	23,204	7,494

Table 2. Characteristics and sources of data for the 33 channels assessed. The Head of Tide (HOT) location is defined as the inland extent of tidal inundation during MHHW.

Name	Free-flowing watershed area upstream of HOT (mi²)	HOT Description	Full Record (calendar years)	Sources of data
Adobe Creek	10.75	SCVWD station 16,745	1980-2012	Sara Duckler (Santa Clara Valley Water District) excel file "SedErosion2002_2012'; Ray Fields (Santa Clara Valley Water District); SFEI excel file "Sediment History 77-04" originally from SCVWD
Alameda Creek	369.02	UPRR Railroad upstream of Ardenwood Blvd	1975-2003	Rohin Saleh (Alameda County Flood Control District); Repeat cross sections and volumes table (Alameda County); Longitudinal profiles (Collins and Leising, 2003); Geomorphic and Sediment Related Studies in the Alameda Creek Flood Control Channel (SFEI, 2012).
Alhambra Creek	15.79	Main Street	2000-2010	Paul Detjens (Contra Costa County Public Works); Joe Enke (City of Martinez); Mark Lindey (ESA/PWA)
Belmont Creek	3.25	Creek crossing with 200 ft downstream of Hwy101	2005-2013	Leticia Alvarez (City of Belmont); Belmont Creek Watershed Study, Creek Assessment, and Recommendations for Sustainable Improvements (WRECO, 2014).
Calabazas Creek	20.32	SCVWD station 2,471	1978-2011	Sara Duckler (Santa Clara Valley Water District) excel file "SedErosion2002_2012'; Ray Fields (Santa Clara Valley Water District).
Colma Creek	12.66	Approximately 1,000ft upstream of Spruce Avenue (near S. Magnolia Ave.)	1997-2013	Julie Casagrande, Mark Chow, and Carole Foster (San Mateo County); Colma Creek Channel Maintenance Project Memo 2 Sediment Processes (Horizon Water and Environment, 2014).
Corte Madera Creek	17.58	Tributary channel just downstream from transition between earthen and concrete channel	1966-2010	Hannah Lee and Davis Hugh (Marin County); Technical Memorandum No.4: Earthen Channel Analysis (Stetson Engineers, 2011a); Captial Improvement Plan (CIP) for Flood Damage Reduction and Creek Management in Flood Zone 9/Ross Valley (Stetson Engineers, 2011b); CMC Analyis of Sediment, April 1984; Stetson Engineers, 2000); Sediment removal PDF (originally from Charlie Goodman) provided by Davis Hugh
Coyote Creek	124.79	SCVWD station number ~55000 (from SFEI HOT report)	2002-2013	Sara Duckler (Santa Clara Valley Water District) excel file "SedErosion2002_2012'; Ray Fields (Santa Clara Valley Water District).
Coyote Creek Marin	1.80	At "neck" (narrowing) just downstream of Ross Drive	2003-2013	Hannah Lee, Joanna Dixon, and Neal Conaster (Marin County); Draft Memorandums 4 and 7 (Noble Consultants, 2013);
Guadalupe River	97.46	SCVWD station ~41,500	1988-2012	Sara Duckler (Santa Clara Valley Water District) excel file "SedErosion2002_2012'; Ray Fields (Santa Clara Valley Water District); SFEI excel file "Sediment History 77-04" originally from SCVWD; Scott Katric (Santa Clara Valley Water District).
Gallinas Creek	1.17	Highway 101	1994-2009	Hannah Lee, Joanna Dixon, and Neal Conaster (Marin County); Channel Maintenance Dredging Las Gallinas Creek San Rafael, CA (Winzler & Kelly, 2010)
Lion Creek	3.31	Footbridge in Coliseum Gardens Park	2007-2013	Moses Tsang (Alameda County Flood Control District); Tom Hinderlie (Alameda County Flood Control District).
Lower Penitencia Creek	29.32	SCVWD station 11,230	1981-2009	Sara Duckler (Santa Clara Valley Water District) excel file "SedErosion2002_2012'; Ray Fields (Santa Clara Valley Water District).

Name	Free-flowing watershed area upstream of HOT (mi²)	HOT Description	Full Record (calendar years)	Sources of data
Matadero Creek	11.47	SCVWD station 10,000	1981-2011	Sara Duckler (Santa Clara Valley Water District) excel file "SedErosion2002_2012'; Ray Fields (Santa Clara Valley Water District); SFEI excel file "Sediment History 77-04" originally from SCVWD.
Napa River	162.31	Just upstream of Trancas	1997-2012	Jeremy Sarrow (Napa County); Jessica Burton Evans and Shelah Sweatt (USACE); HEC model cross sections vs 2012 dredge survey cross sections (West Consultants 2014); EIS/EIR report (DCE, 1999); CESPK-ED-D Memo for File (History of Napa Dredging); USACE DMMO Annual Report.
Novato Creek	14.83	Warner Creek confluence with Novato Creek	1983-2012	Hannah Lee, Pat Valderama, and Joanna Dixon (Marin County); Sediment Sources and Fluvial Geomorphic Processes of Lower Novato Creek Watershed (Laurel Collins, 1998); Flood and Sediment Study for Lower Novato Creek (PWA, 2002); Hydraulic Assessment of Existing Conditions Novato Creek Watershd Project (KHE, 2014).
Old Alameda Creek	20.36	Hesperian Blvd	2000-2013	Moses Tsang (Alameda County Flood Control District).
Permanente Creek	16.65	SCVWD station 8,700	1979-2009	Sara Duckler (Santa Clara Valley Water District) excel file "SedErosion2002_2012'; Ray Fields (Santa Clara Valley Water District); SFEI excel file "Sediment History 77-04" originally from SCVWD.
Petaluma River	44.63	Near Lakeville Street	2003-2013	Jon Niehaus (Sonoma County Water Authority); Pam Tuft (City of Petaluma); Jessica Burton Evans and Shelah Sweatt (USACE); USACE DMMO reports.
Pinole Creek	14.43	60 m downstream of railroad trestle	1965-2010	Paul Detjens (Contra Costa County Public Works); Rich Walkling (Restoration Design Group); Sarah Pearce (SFEI); Pinole Creek Watershed Sediment Source Assessment, (Pearce et al., 2005).
Rodeo Creek	10.12	4th street	1993-2013	Paul Detjens (Contra Costa County Public Works); Rodeo Creek Vision (Restoration Design Group, 2008); Restoration Design Group unpublished cross sections 2006; Stream Network and Landscape Change in the Rodeo Creek Watershed (Collins, 2008).
San Bruno Creek	4.52	Tide gate, very near mouth of channel	2008-2013	Julie Casagrande, Mark Chow, and Carole Foster (San Mateo County); Nixon Lam (San Francisco International Airport, Planning and Environmental Affairs)
San Francisquito Creek	30.64	Highway 101	1958-2007	Kevin Murray (San Francisquito Joint Powers Authority); Watershed Analysis and Sediment Reduction Plan (NHC, 2004); San Francisquito Creek Flood Damage Reduction & Ecosystem Restoration Feasibility Report (NHC, 2010); Sara Duckler (Santa Clara Valley Water District) excel file "SedErosion2002_2012'.
San Leandro Creek	6.35	Railroad grade on downstream side of Highway 880	2000-2013	Moses Tsang (Alameda County Flood Control District).
San Lorenzo Creek	21.92	End of concrete channel, just upstream of Railroad Ave	2002-2013	Moses Tsang (Alameda County Flood Control District); Arthur Valderrama (Land Development, Alameda County Public Works); San Lorenzo Creek Bulk Sediment Study (unpublished SFEI, 2012).
San Pablo Creek	9.08	Fred Jackson Way	2003-2013	Paul Detjens (Contra Costa County Public Works)

Name	Free-flowing watershed area upstream of HOT (mi²)	HOT Description	Full Record (calendar years)	Sources of data
San Tomas Aquino Creek	41.56	SCVWD station 15,690	1977-2012	Sara Duckler (Santa Clara Valley Water District) excel file "SedErosion2002_2012'; Ray Fields (Santa Clara Valley Water District).
Sonoma Creek	92.16	Highway 12/El Camino Real	2000-2013	Jon Niehaus (Sonoma County Water Authority); Sonoma Creek TMDL; Sediment Source Assessment, Sonoma Creek Watershed (Rebecca Lawton, 2006) http://knowledge.sonomacreek.net/SSA.
Stevens Creek	12.76	SCVWD station 11,122	1980-2012	Sara Duckler (Santa Clara Valley Water District) excel file "SedErosion2002_2012'; Ray Fields (Santa Clara Valley Water District); SFEI excel file "Sediment History 77-04" originally from SCVWD.
Sunnyvale East Channel	6.63	SCVWD station 11,184	1979-2006	Sara Duckler (Santa Clara Valley Water District) excel file "SedErosion2002_2012'; Ray Fields (Santa Clara Valley Water District).
Sunnyvale West Channel	2.82	SCVWD station 14,718	1981-2011	Sara Duckler (Santa Clara Valley Water District) excel file "SedErosion2002_2012'; Ray Fields (Santa Clara Valley Water District).
Walnut Creek	122.21	Weir just downstream of Highway 4	1965-2007	Paul Detjens (Contra Costa County Public Works); Walnut Creek Sedimentation Report (MBH, 2012); Sediment composition study (Teeter, 2010).
Wildcat Creek	8.12	Just downstream of Richmond Parkway	1989-2011	Paul Detjens (Contra Costa County Public Works); "WC Silt Hist" table from Paul Detjens; Wildcat Creek Watershed Study (Collins et al., 2001).

ABAG (Association of Bay Area Governments). 2006. Existing land use in 2005: Data for Bay Area Counties. Association of Bay Area Governments. https://store.abag.ca.gov/projections.asp#elu

BCDC (San Francisco Bay Conservation and Development Commission). 2013. Corte Madera Baylands Conceptual Sea Level Rise Adaptation Strategy Report. Prepared by: The San Francisco Bay Conservation and Development Commission and ESA PWA, May 2013. http://www.bcdc.ca.gov/climate_change/WetlandAdapt.html

Beyeler, M., Mehaffy, M., Connor, M., Doehring, C., Lowe, J., Grossinger, R., Senn, D., and Novick, E., 2015. Decentralized Wastewater Discharges and Multiple Benefit Natural Infrastructure: Preliminary Analysis and Next Steps (Final project report). Report prepared for the East Bay Dischargers Authority (EBDA). 117 pp. http://www.sfei.org/ebda-sea-level-plan#sthash.bd4ARQuA.dpbs

Collins, L. 1998. Sediment Sources and Fluvial Geomorphic Processes of Lower Novato Creek Watershed. Technical report prepared for the Marin County Flood Control and Water Conservation District.

Collins, L., J. Collins, R. Grossinger, L. McKee, and A. Riley. 2001. Wildcat Creek Watershed: A Scientific Study of Physical Processes and Land Use Effects. Technical report funded by the Lucille and David Packard Foundation and the Contra Costa Clean Water Program, Publication #363, San Francisco Estuary Institute, Richmond, CA.

Collins, L., and K. Leising. 2003. Draft Alameda Creek Sediment Budget and Sediment Source Analysis: 2003 Progress Report. Technical report prepared for Alameda County Public Works Department.

Collins, L. 2006. Estimating Sediment Supplies from Local Watersheds to Intertidal Habitats of San Francisco Bay. Paper given at the 2006 South Bay Science Symposium for the South Bay Salt Pond Restoration Project, San Jose, CA.

Collins, L., 2008. Stream Network and Landscape Change in the Rodeo Creek Watershed. Technical report prepared for the Muir Heritage Land Trust.

DCE (Design, Community and Environment). 1999. Napa River/Napa Creek Flood Reduction Project: Final Supplemental Environmental Impact Statement/ Environmental Impact Report. Prepared for the US Army Corps of Engineers and the Napa County Flood Control and Water Conservation District. Prepared by Design, Community and Environment, Berkeley, CA. 920 pp.

Goals Project. 2015. The Baylands and Climate Change: What We Can Do. Baylands Ecosystem Habitat Goals Science Update 2015 prepared by the San Francisco Bay Area Wetlands Ecosystem Goals Project. California State Coastal Conservancy, Oakland, CA.

Golden Gate Weather Services. 2016. San Francisco Seasonal Rainfall: 1849-1850 to 2008-2009. http://ggweather.com/sf. Accessed on 10/1/2016.

Healey, M.C., Angermeier, P.L., Cummins, K.W., Dunne, T., Kimmerer, W.J., Kondolf, G.M., Moyle, P.B., Murphy, D.D., Patten, D.T., Reed, D.J. and Spies, R.B., 2004. Conceptual Models and Adaptive Management in Ecological Restoration: The CALFED Bay-Delta Environmental Restoration Program.

Horizon Water and Environment. 2014. Colma Creek Channel Maintenance Project. Technical report prepared for San Mateo County Department of Public Works. Prepared by Horizon Water and Environment, Oakland, CA.

Jennings, C.W., Strand, R.G. and Boylaw, R.T., 1977. Geologic Map of California. Division of Mines and Geology.

KHE (Kamman Hydrology and Engineering, Inc). 2014. Hydraulic Assessment of Existing Conditions, Novato Creek Watershed Project. Technical report prepared for Marin County Department of Public Works. Prepared by Kamman Hydrology and Engineering, Inc., San Rafael, CA. 198 pp.

Lawton, R. 2006. Sonoma Creek Watershed Sediment Source Assessment. Technical report prepared by Sonoma Ecology Center with contributions by Watershed Sciences, Martin Trso, Talon Associates, and Tessera Consulting. 282 pp.

MBH (Mobile Boundary Hydraulics). 2012. Walnut Creek Sedimentation Study. Technical report prepared by R. Copeland for the U.S. Army Corp of Engineers Sacramento District, Sacramento, CA.

McKee, L.J., Lewicki, M., Schoellhamer, D.H. and Ganju, N.K., 2013. Comparison of Sediment Supply to San Francisco Bay from Watersheds Draining the Bay Area and the Central Valley of California. Marine Geology, 345, pp.47-62.

NAIP. 2012. United States Department of Agriculture, National Agriculture Inventory Program. https://www.fsa.usda.gov/programs-and-services/aerial-photography/imagery-programs/naip-imagery/

Noble Consultants, Inc. 2013. Coyote Creek (Marin) Sediment Stabilization Project. Technical report prepared for Marin County Flood Control and Water Conservation District. Prepared by Noble Consultants, Novato, CA. 10 pp.

NHC (Northwest Hydraulic Consultants Inc). 2004. San Francisquito Creek Watershed Analysis and Sediment Reduction Plan. Prepared for San Francisquito Creek Joint Powers Authority. Prepared by Northwest Hydraulic Consultants, West Sacramento, CA. 128 pp.

NHC (Northwest Hydraulic Consultants Inc). 2010. San Francisquito Creek Flood Damage Reduction and Ecosystem Restoration Feasibility Report. Prepared for San Francisquito Creek Joint Powers Authority. Prepared by Northwest Hydraulic Consultants Inc., West Sacramento, CA. 57 pp.

NRC (National Research Council). 2012. Sea-Level Rise for the Coasts of California, Oregon, and Washington: Past, Present, and Future. Washington, DC: The National Academies Press.

Pearce, S., L. McKee, and S. Shonkoff. 2005. Pinole Creek Watershed Sediment Source Assessment. Technical report of the Regional Watershed Program, Publication #316, San Francisco Estuary Institute, Oakland, CA.

PWA (Philip Williams & Associates). 2002. Flood and Sediment Study for Lower Novato Creek. Technical report prepared for Marin County Flood Control and Water Conservation District. Prepared by Philip Williams and Associates, San Francisco, CA. 20 pp.

RDG (Restoration Design Group). 2008. Rodeo Creek Vision Plan. Prepared for Contra Costa County Flood Control and Water Conservation District. 24 pp.

SFEI (San Francisco Estuary Institute). 1998. Bay Area EcoAtlas: Geographic Information System of Wetland Habitats Past and Present. http://www.sfei.org/ecoatlas/.

SFEI (San Francisco Estuary Institute). 2012. Geomorphic and Sediment Related Studies in Alameda Creek Flood Control Channel: Synthesis of Recent Scientific and Engineering Studies and Options for Consideration. Prepared by SFEI with contributions from Restoration Design Group (Roger Leventhal and Erik Stromberg), Watershed Sciences (Laurel Collins), Center for Ecosystem Management and Restoration (Matthew Deitch and Gordon Becker), Bigelow Watershed Geomorphology (Paul Bigelow), and Swanson Hydrology and Geomorphology (Mitchell Swanson). A technical report prepared for the Alameda County Flood Control and Water Conservation District (ACFC&WCD) by the Regional Watershed Program, San Francisco Estuary Institute, Richmond, CA.

SFEI (San Francisco Estuary Institute). 2014. Bay Area Aquatic Resource Inventory. http://www.sfei.org/BAARI. [GIS layers]

SFEI-ASC (San Francisco Estuary Institute-Aquatic Science Center). 2015. Novato Creek Baylands Vision: Integrating Ecological Functions and Flood Protection Within a Climate-Resilient Landscape. A SFEI-ASC Resilient Landscape Program report developed in cooperation with the Flood Control 2.0 project Regional Science Advisors and Marin County Department of Public Works, Publication #764, San Francisco Estuary Institute-Aquatic Science Center, Richmond, CA.

SFEI-ASC (San Francisco Estuary Institute-Aquatic Science Center). 2016. Resilient Landscape Vision for lower Walnut Creek: Baseline Information & Management Strategies. A SFEI-ASC Resilient Landscape Program report developed in cooperation with the Flood Control 2.0 Regional Science Advisors and Contra Costa County Flood Control and Water Conservation District, Publication #782, San Francisco Estuary Institute-Aquatic Science Center, Richmond, CA.

Stetson Engineers. 2000. Geomorphic Assessment of the Corte Madera Watershed. Technical report prepared for Friends of Corte Madera Creek Watershed and the Marin County Department of Public Works. Prepared by Stetson Engineering, San Rafael, CA. 91 pp.

Stetson Engineers, 2011a. Technical Memorandum No. 4: Earthen Channel Analysis. Prepared for Marin County Flood Control and Water Conservation District, Flood Zone 9. Prepared by Stetson Engineers, San Rafael, CA. 22 pp.

Stetson Engineers. 2011b. Capital Improvement Plan (CIP) for Flood Damage Reduction and Creek Management in Flood Zone 9/Ross Valley. Prepared for Marin County Flood Control and Water Conservation District, Flood Zone 9. Prepared by Stetson Engineers, San Rafael, CA. 50 pp.

Teeter, A. 2010. Walnut Creek HEC-6 Modeling- Sediment Data Analysis and Suggested Sediment Model Parameters. Technical report prepared for Contra Costa County Flood Control and Water Conservation District. 27 pp.

USACE Dredge Material Management Office (DMMO) Annual Reports. http://www.spn.usace.army.mil/Missions/Dredging-Work-Permits/Dredged-Material-Management-Office-DMMO/Annual-Reports/

West Consultants. 2014. Napa River Dredging Survey vs Comparison. Technical Memo prepared for the Napa County Flood Control and Water Conservation District. Prepared by West Consultants, Inc., Folsom, CA. 18 pp.

Winzler and Kelly. 2010. Channel Maintenance Dredging Las Gallinas Creek San Rafael, CA. Technical report prepared for the Marin County Department of Public Works. Prepared by Winzler and Kelly, Santa Rosa, CA. 98 pp.

Witter, R.C., Knudsen, K.L., Sowers, J.M., Wentworth, C.M., Koehler, R.D., Randolph, C.E., Brooks, S.K. and Gans, K.D., 2006. Maps of Quaternary Deposits and Liquefaction Susceptibility in the Central San Francisco Bay Region, California (No. 2006-1037). Geological Survey (US).

WRECO. 2014. Belmont Creek Watershed Study, Creek Assessment, and Recommendations for Sustainable Improvements, San Mateo County, California. Prepared for Novartis Pharmaceuticals Corporation. 144 pp.